Foraging in 2020:

The Ultimate Guide to Foraging and Preparing Edible Wild Plants With Over 50 Plant Based Recipes

Joseph Erickson

© Copyright 2020 by Joseph Erickson. All right reserved.

The work contained herein has been produced with the intent to provide relevant knowledge and information on the topic on the topic described in the title for entertainment purposes only. While the author has gone to every extent to furnish up to date and true information, no claims can be made as to its accuracy or validity as the author has made no claims to be an expert on this topic. Notwithstanding, the reader is asked to do their own research and consult any subject matter experts they deem necessary to ensure the quality and accuracy of the material presented herein.

This statement is legally binding as deemed by the Committee of Publishers Association and the American Bar Association for the territory of the United States. Other jurisdictions may apply their own legal statutes. Any reproduction, transmission or copying of this material contained in this work without the express written consent of the copyright holder shall be deemed as a copyright violation as per the current legislation in force on the date of publishing and subsequent time thereafter. All additional works derived from this material may be claimed by the holder of this copyright.

The data, depictions, events, descriptions and all other information forthwith are considered to be true, fair and accurate unless the work is expressly described as a work of fiction. Regardless of the nature of this work, the Publisher is exempt from any responsibility of actions taken by the reader in conjunction with this work. The Publisher acknowledges that the reader acts of their own accord and releases the author and Publisher of any responsibility for the observance of tips, advice, counsel, strategies and techniques that may be offered in this volume.

TABLE OF CONTENTS

INTRODUCTION 1

CHAPTER 1 *WHAT IS FORAGING?* 4

    WHAT DOES IT MEAN TO FORAGE? 4

    HISTORY OF FORAGING 5

    FORAGING NOW 6

    FORAGING SKILLS 8

    FORAGING SAFETY 8

    FORAGING BENEFITS 9

    FORAGING HAZARDS 11

    KEEP YOUR EYES OPEN 13

    HOW TO FORAGE RESPONSIBLY 15

    DON'T GET LOST 18

    DON'T PANIC 18

    FIGURE OUT IF YOU WILL MOVE OR STAY 20

    BE READY FOR ANYTHING 21

CHAPTER 2 *WHERE TO FORAGE?* 23

    DON'T HARM YOURSELF 26

    CONTAMINATION 27

    HOW TO FIND WILD EDIBLES IN A CITY 28

    URBAN PLANTS YOU CAN EAT 29

    THERE'S A MAP FOR THAT 31

CHAPTER 3 *WHEN TO FORAGE?* 33

    SPRINGTIME FORAGING 34

    Summertime Foraging 37

    Autumn Foraging 39

    Winter Foraging 42

    Preserving Your Food 45

CHAPTER 4 *TOOLS FOR FORAGING* **48**

    Ethical Foraging 55

CHAPTER 5 *IDENTIFYING PLANTS* **60**

    How to Identify Edible Plants 61

    Know Before You Go 63

    In the Woods 64

    What Not to Eat 64

    Identifying Poisonous Plants 69

    Plants in Your Yard 70

    You Touched a Poisonous Plant, Now What? 70

CHAPTER 6 *LIST OF EDIBLE WILD PLANTS* **71**

CHAPTER 7 *LIST OF MEDICINAL WILD PLANTS* **82**

CHAPTER 8 *LIST OF POISONOUS PLANTS* **92**

CHAPTER 9 *RECIPES FOR FORAGED PLANTS* **102**

    Stinging Nettle Spanakopita 102

    Dandelion Fritters 105

    Seaweed Salad 106

    Wild Mushroom Ragu 107

    Stir-Fried Dandelion Greens 109

    Purslane Tacos 110

    Zucchini and Purslane Soup 112

- Purslane Salad 114
- Stinging Nettle Soup 115
- Creamy Nettle Soup 117
- Fiddlehead Soup 119
- Dandelion and Violet Lemonade 121
- Ginger, Pineapple, and Purslane Smoothie 122
- Purple Dead Nettle Tea 123
- Violet Syrup 124
- Wild Violet Vinegar 125
- Dandelion Root Chai 126
- Fireweed Tea 128
- Hibiscus Syrup 129
- Wild Ramp Pesto 130
- Polish Fermented Mushrooms 132
- Garlic Mustard Pesto 134
- Sorrel Sauce 135
- Fennel Sauerkraut 136
- Country Mustard 137
- Ancient Roman Mustard 138
- Pickled Blueberries 139
- Pickled Fiddleheads 140
- Candied Angelica 141
- Strawberry Dandelion Cake 142
- Douglas Fir Poached Pear and Frangipane Tart 145
- Wintergreen Ice Cream 148
- Mulberry Sorbet 150
- Gooseberry Sorbet 151

Wild Cranberry Sauce *152*

Paw Paw Ice Cream *153*

Black Walnut Snowball Cookies *154*

CONCLUSION **155**

INTRODUCTION

First off, I would like to thank you for choosing this book, and I hope that you find it informative and helpful no matter what your goals may be. Throughout this book, we will discuss what foraging is and how you can safely forage for some of your food.

Foraging is something that most people have heard of. We learned in history class that our ancestors, the hunter-gatherers, were foragers. That was their only means of finding food. But then times changed and grocery stores became our main source of produce. Sure, they are convenient, and you can find food there that you would never find in your backyard, but they will never be as fresh as what you can forage.

A person who forages for part of their food will become more independent in their food supply. They won't be handing over as much of their money to large corporations. They also know exactly where their food came from. When shopping at the grocery store, you don't know as much about the produce. You aren't sure how long it has been since they were picked, or how long they have been sitting on the shelf. Plants are always better when eaten as close to picking as possible. They have a higher nutrient content at that point. But most of the food at the store has traveled hundreds of miles, and they were picked before they were ever fully ripe. Plus, if you want to make sure you aren't consuming any chemicals, you are going to have to get organic produce, which can get very pricey.

A better alternative is learning how to forage, which is what your ancestors would have done. Foraging can sound scary or even weird, especially in this day and age. That's why you have this book, though. It can help guide you through the process, so you know what to do and how to do it safely.

We will go over everything you need to know. We'll discuss some of the best places to forage for food, even if you live in an urban setting. We'll also go over what you should look for during certain times of the year, and contrary to what some people may believe, you can still forage during the winter. You'll find all of the tools that you will need for foraging as well.

Of course, we will go over how to identify plants so that you can safely forage safe foods that won't end up harming you. We even have a chapter devoted to talking about plants that are poisonous to humans so that you never eat something that you shouldn't.

The great thing is, a lot of these plants that you will be foraging for also have medicinal properties. Not only can they make a delicious dish, but they can also be turned into medical products that can help heal different ailments. Of course, you should not allow them to replace a regular doctor, only as a supplement to minor illnesses.

Foraging is a lot of work, but it is rewarding work. And you will be amazed at how many edible plants are in your backyard that you may have been mowing down for years. Adding foraging to your routine can help you save money, eat healthier, and feel better about your impact on the Earth.

As long as you refer back to this information often, and use common sense, you can safely forage for food without harming yourself or the planet.

Before we get started, I would like to ask that if you find any part of this book helpful, please leave a review. Again, thank you for choosing this book.

CHAPTER 1

What Is Foraging?

Foraging for food in the wild has become more popular in the past few years. This has become a trendy activity for some people but a total lifestyle for others due to the awareness about eating fresh, local, organic ingredients.

What Does It Mean to Forage?

Foraging normally refers to looking for, identifying, and then collecting food from the wild like shellfish, mushrooms, nuts, fruits, plants, and herbs. It is about preserving, cooking, and eating nutritious food that is healthy while understanding natural resources.

Most people have fond memories of picking blackberries with the family. This is a perfect example of foraging that many people have enjoyed at one time or another during their life. Elderflowers, sloes, and apples have also been picked in the wild to make delicious preserves in our homes.

Ireland and the United Kingdom have been blessed with a huge range of edible plants, fruits, and mushrooms that grow near the villages plus in the wild. You'll be surprised to find that a lot of the plants will grow in places you frequent, your backyard, or your local park.

Cherry trees that have been planted in parks, blackberry brambles like taking over unused lots, and hazel trees that line the street drop lots of nuts that are waiting to be picked up. By having a good knowledge of certain plants could help you find precious edibles like some wild mushrooms.

Any of these ingredients could be used to cook stews and soups or preserved as chutneys and jams. You could even make your own liquors, wines, or teas to get their medicinal properties.

History of Foraging

Foraging has been around since the beginning of the human race. The hunter-gatherers would sustain their tribes by gathering plants in their environment. They would also hunt animals. This continued until they could invent agriculture and farming as a better source of food. All the food that they ate had to be gathered by their own hands like mushrooms from trees that are decaying, wild greens from fields, and berries found on bushes. With a lot of practice, they learned how to identify the plants that are edible, poisonous, and which ones can be found at various times throughout the year.

Foraging played a huge role during hard times that were brought on by hunger and poverty. During World War II, rosehips gave soldiers vitamin C when there was a shortage of oranges, and dandelion root and acorns were used as a substitute for coffee.

Households would supplement their pantry with medicinal plants and food that were gathered in the wild until the first supermarkets were built. Even though gathering food in the wild is still part of daily life for many who live in rural areas in Europe, having a good knowledge of plants that grow in the wild has been forgotten.

Foraging Now

The food system today is a lot different. In the United States, most of our food is grown on large farms that aren't anywhere close to the people that will eat the food. For most people, "gathering food" simply means taking a trip to their local supermarket. Most probably haven't seen food growing out of the ground. Until we bring it into our houses, we don't have any connection to our food. We are limiting our choices to unhealthy foods that don't give us any valuable nutrients.

It doesn't need to be this way. In this modern world of ours, it is possible to find and pick some wild plants that aren't just edible but very nutritious. It can be a bit risky since we don't possess the knowledge of our ancestors. If a person chooses to march into the woods and starts munching on the first thing they find, they could end up facing a horrible tummy ache or worse.

It can take you years to be able to know every wild plant that could be eaten. What is more important is you have to recognize the plants that you absolutely should not eat. There are common plants that beginners can easily identify and enjoy. Adding some dandelion greens to your salad or picking some wild raspberries or blackberries while hiking will give you a chance to enjoy some fresh food that is free and delicious. It also connects you to nature and your ancestors. Plus, there is a thrill when you eat things that you have found and picked with your hands that could give you an appetite for learning more about all that nature can give us.

Naturally occurring foods are healthy was to supplement our diets. This isn't the only reason why people like to forage and take things back to the old wisdom and ways. There does seem to be an awareness about where the foods we eat actually come from.

Due to some recent scandals within the food industry, raising the awareness for leaving carbon footprints and helping the environment has caused many people to think about different ways to find our food, and this, in turn, will bring us closer to nature.

Foraging doesn't apply exclusively to the people who live in the country, but it also includes finding foods on the side of the road, in verges, and parks in a large city. Most of the time, parks will have plants that you won't ever find growing in the wild, and this gives us some very interesting finds.

Wildlife tourism has attracted a lot of people to begin foraging as a recreational hobby. This allows you to get away from our daily lives and find new and amazing ingredients that taste great and will enhance our meals.

Eventually, for some people, foraging is healthy and will help empower us. It doesn't just help us satisfy our nutritional needs but can help us understand the way nature works. Foraging can totally change how you see your surroundings and the world.

Foraging Skills

Foraging hasn't changed a lot since the beginning of humans. There are leafy plants, roots, berries, mushrooms, medicinal herbs, and nuts that are very nutritious and edible. Foraging is a skill that has been forgotten by most people. Our ancestors might have known about harvesting dandelions to make salads and wine, and they may have known that elderberries can be used to make cough syrup and wine, but most people who live in cities of all sizes have become distanced from the plants and the knowledge of these plants. Foraging skills have almost been completely lost.

Foraging Safety

While teaching a children's survival class one day, the subject of how dangerous it was to eat wild plants came up. One parent spoke up and stated that while he was an Army Ranger, he had to take many soldiers off the line because they got sick from eating something they found in the wild.

If a person is hungry enough and they see a plant full of berries, it would be natural to think that you can stay alive by eating them. If we had to choose to eat a handful of insects or a handful of greens, we are going to choose to eat the greens. The biggest problem with that is only about five to ten percent of all plants will be edible. The rest will either be unpalatable or poisonous.
This is why you have to know ways to forage safely.

You need to find a foraging guide that has good pictures and very well detailed descriptions of all the edible plants. All the descriptions need to have information about any look-alike plants that might be deadly. Elderberries are great to be used medicinally and are delicious to eat. The plant and berry can be confused with pokeberries and water hemlock. A good app, website, or book will list all the dangerous look-alikes along with the potential dangers of all edible plants.

Foraging Benefits

Finding your own food in the wild is a lot more than just being fun, even though it is that. Gathering food gives you many benefits like these:

- Connecting to Nature

Finding your own food can restore your connection to nature and the seasons that have been lost in today's world. You could watch the progression of spring by looking for the arrival of every new plant like morel mushrooms, ramps, watercress, and dandelion greens. In the summertime, you can look for ripening persimmons, wild raspberries, wild blackberries, and juneberries. When you gather these wonderfully delicious plants, you will feel like you are an actual part of nature and not just a mere observer.

- Sustainability

Any plant that you pick yourself is locally grown and organic. These plants haven't been grown with agricultural chemicals or harmful pesticides. The only water they have been exposed to is rain. They don't require fossil fuels to be harvested and transported to stores. When you go walking through the forest and gather some wild greens, they won't have any carbon footprint.

- Exercise

Foraging for food in the great outdoors will get you outside and moving. Hiking to your favorite harvesting spot, reaching and stretching to pick berries, and bending to harvest greens can be a great workout. It will be more relaxing and pleasant than spending an hour at your local gym walking on a treadmill under harsh fluorescent lights.

- Nutrition

For the most part, plants that you forage for will be more nutritious than their grocery store counterparts. The main reason why is because you get to consume them closer to harvest when they have their maximum nutrients. Wild dandelion greens will have seven times more phytonutrients than the spinach you will find in the grocery store. You can find edible crab apples that could have between two to 100 times the phytonutrients that are found in the apples at the grocery store. One good bonus is when you are outside picking your plants; you are giving your body a lot of vitamin D.

- New Flavors

Some wild foods are a bit hard to find in regular grocery stores like wild mushrooms. There are some that won't ever find in supermarkets like pawpaws. This is a fruit similar to the mango that has a custardy texture that is way too delicate to ship to stores. Foraging is usually your best chance to find these tasty and unique foods.

- Food for Free

When you forage for food, it doesn't cost anything except the time it takes you to find and pick it. Yes, your time is worth something, and it might not be worth your effort to forage foods such as potatoes. Most of the foods that can be found in the wild cost a lot of money when found in the grocery store. Pine nuts, ramps, and chanterelle mushrooms can cast more than $20 per pound, if not more.

Foraging Hazards

Even though foraging for food gives you all the benefits above, it does have some pitfalls, too, especially if the person foraging is a beginner. Some risks of foraging might include:

- Being Arrested

It is illegal for you to forage on another person's property if you don't have their permission. Many state and federal parks forbid you from gathering plants unless you are lost, and you don't have any supplies, and you are just trying to survive. In places where it is legal to forage, there are limits on the number of plants you can take. Just to make even more complicated, boundaries won't be marked clearly, so it isn't easy to know if you have wandered into a place where foraging has been banned. People who just walk into a field and begin grabbing plants might find themselves in handcuffs.

- Damage to the Environment

It isn't obvious how much of one plant you can take without killing it completely. Foragers who are overenthusiastic might strip all of the native plants out of one area. This could create an opening for invasive plants to grow and take over. Foragers could damage a delicate environment just by walking on it. They damage the topsoil, they crush plants, and they disrupt the habits of wildlife. Speaking with a seasoned forager, naturalist, or game warden can help you learn more about which areas are too vulnerable to be disturbed.

- Handling Unfamiliar Foods

Foragers who are experienced know which plants can be eaten, but they also need to know how to eat them. Most of the wild, edible plants are indigestible, bitter, or tough. If you don't know how to prepare them the right way, they aren't going to be edible. Some inexperienced foragers might find themselves looking at a basket of wild produce going to waste because they don't know how to cook with it.

- Consuming Harmful Plants

Even though most wild plants are nutritious and tasty, some might be poisonous, and most of them look like plants that are edible. Beginning foragers are most likely to mistake a plant that is poisonous for one that is edible. Christopher McCandless wrote a book on edible wild plants, but the book was published posthumously after he ate a poisonous plant in the Alaskan wilderness and died. The book is called *Back to the Wild*. Even if you are positive that the plant isn't toxic, it could make you sick if it was contaminated by chemicals, animal waste, or pesticide residue.

Keep Your Eyes Open

The biggest rule when foraging is to keep your eyes open. You might be surprised to find new foods or plants every time you look at the ground. The biggest appeal for foragers is finding, identifying, and harvesting foods that grow in the wild without any help from humans. If you are armed with some information, you can venture into a forest close to you, city part, and even a parking lot and bring dinner home. Just remember to follow all local park laws about picking plants.

- Chefs and Foraged Foods

Chefs at high-end restaurants are getting interested in foraged foods, and this has given foraged foods a boost recently. They like foraged foods because it enhances their dishes. It can also help them supplement their menus by using local ingredients that make their dishes stand out. Farmer's markets are now being filled with delicacies such as morel mushrooms and ramps along with other edible plants like lamb's quarters, nettles, and dandelion. Chefs have been working hard to get the word spread about the easiness of finding these goodies in your own backyard.

- Foraging Begins with Learning

Foraging seems like it would be an easy solution for a lot of problems with our food system, like access, nutrition, and sustainability. Eating wild is an easy way to eat sustainable if you do it responsibly. Since foraging is free, anybody is able to do it. It also provides urban dwellers access to ingredients that are more nutritious, better quality, and tastier than why they have available. Why is it so hard to get people to understand how these foods can benefit them?

One big problem is most people don't like the thoughts of eating things straight from the ground. There is an "ick factor" that some people just can get over, but with some education about the plants that are edible and the location they grow, can fix this. Most people will look at these kinds of foods as "starvation" foods. They would only eat them if it was their last resort to keep them from starving. There are many different reactions from: "I won't eat anything that has been picked from the dirt." or "this is very empowering." If these people would just stop and think about the foods that are on the grocery store shelves, they have come from the ground somewhere.

- Foraging Regulations

Not having enough information isn't just a top-down problem. Local governments still have a lot to learn when it comes to foraging being viable. This is why Berkeley Open Source Food introduced some policies to legislators on ways that foraging could supplement government assistance. One article advocated for governments to open public lands for forging and would give training to help people who live in areas where food isn't readily available, ways to identify, and then harvest the plants correctly. People look at these areas as "food deserts," but if plants grow there and they are safe to eat, this it really isn't a food desert; it is actually an information desert.

Head chef at Larder, located in Cleveland, Jeremy Umansky, is a licensed mushroom hunter who says foraging helps bring more food to his restaurant. Foraged plants make up a big percentage of his menu.

He states that anybody can forage. When you decide to process, sell, or serve the foods you find, that's when you have to have a license. This isn't saying that just because you don't have a license, you don't know what you are doing. You might have a person who never went to college, but they are just as good of a cook as the girl next door.

How to Forage Responsibly

Leda Meredith leads foraging tours in New York and wrote several books about this subject. She has created the best practices, and help come up with a bibliography for anybody who has the courage to forage.

- You Have to Know Your Plants

Beginners to foraging need to take time to become familiar with the plants. Find a guide, go on some tours, or take a class or two. You begin with, "it has this kind of leaf, and its flower is this color," and then go down the list and make sure you can check everything off of it.

Having some knowledge of your local plants will go a long way toward getting rid of any misconceptions. People are super afraid of mushrooms because there are so many that can look like edible ones but aren't. Mushrooms have to be eaten to have any negative side effects. You might touch one of the most toxic mushrooms, but if you don't eat it, it isn't going to do anything to you. On the other hand, plants have sap that could get on your skin and cause all sorts of problems, and some even have burrs and thorns. Anybody who has brushed up against stinging nettles knows exactly what I'm saying, but in all these cases, a bit of knowledge will go a long way.

Just a reminder, if you were to miss-identify a mushroom and consume it, it can end up being a deadly mistake. One good rule of thumb is if you are not 100 percent sure about the identity of a mushroom, don't eat it.

- Know the Soil

There is a misconception that urban soil has been laced with all kinds of contaminants, including high levels of PCBs, pesticides, and herbicides. This causes some foragers to worry that the food they find might harm their health. You can learn more about your local soil by taking a sample of it and getting it analyzed by your local county office. This may cost a bit of money, but it could be worth it. In one study, soil samples were taken from three different spots in the Bay Area. Some of the tested soil had high levels of cadmium and lead, but not of it had been transferred into the plants in that area.

None of the plant's tissues were accumulating any heavy metals and were completely safe to eat. The typical plant that grows wild has built up some resistance to harsh environments and are sometimes more nutritious than what is being farmed and transported to the stores. They have developed their own mechanisms to detox.

If you don't have any access to any testing equipment, you can make a phone call to your city manager and talk with the park's department. They can tell you if they use agrochemicals in your area or if it is a parcel that was reclaimed.

- Using a Newborn Stomach

When you finally identify a wild plant, for the very first time, it can be very exciting, but you have to have some restraints. If you are foraging for foods that you haven't ever eaten before, you will want to treat your body the same way that you treat a newborn. Try a few ounces of a plant and then wait for 24 hours.

Other than making sure you take precautionary measures, you should make sure you take things slow as it can have some positive environmental consequences. For example, if you make sure you don't just shove a bunch of berries in your mouth, and instead take the time to think about how you want to use and store it, you won't overharvest from a plant.

- Never Overharvest

Other than protecting your health and taking it easy on your digestive system, having some restraint is the main rule when foraging. Think about the land where you are foraging. Never take more than you will need or more than you will use.

There are thousands of medicinal or edible plants that are thought to be invasive weeds. Plants like Japanese knotweed, burdock, and dandelion, like being bullies and crowding out all the other plants around it. You can harvest them freely without damaging the ecosystem. You will actually be helping all the native plants.

In the case of the market favorite ramps, you have to forage the plant correctly and sparingly so that there s enough left for the next season. Once ramps are in season, only cut the leaves in spring and don't dig the bulbs until July. Every part of the plant will have a different season, but if you correctly treat the plant, you can enjoy it through many seasons.

Don't Get Lost

Anybody could get lost in the woods. You might be an experienced backpacker or just a day hiker, but one wrong move might send you to a place where you might be forced to rely on your own skills and very limited supplies. If you go into the woods with the correct tools, a good plan, and some know-how, you will be able to get back to civilization with poise, and it won't matter where you go.

Don't Panic

If you realize that sunset is approaching fast and your go around a clump of trees thinking that the trailhead was just on the other side of it, but you don't see it anywhere, your first reaction might be to panic. For many people, just thinking about being lost might be enough to trigger your anxiety, but the main thing is you have to stay calm. When you made a decision out of fear, these won't ever be constructive, and any drastic measure that was taken without thinking about it could make the bad situation a lot worse.

Rather than panicking, you need to STOP:

- S: stop walking; if you continue to wander around, you might get even more lost.
- T: think objectively, so you will be able to see the situation better and avoid panic.
- O: observe your surroundings. This includes any immediate risks, all available resources, and the weather.
- P: plan to handle the situation when you have the complete picture.

Remember that not all wooded threats have been created equal. Your most dangerous one would be the lack of oxygen that could be caused by a cave-in or avalanche. The second one would be being exposed to temperatures that are warmer or colder than what your body is able to stand. Not having enough water comes in third, and the last is not having enough food. You have to look at and address every risk in the order of their severity to help increase your chances of surviving if a serious condition arises.

The main factor in deciding to stay where you are and waiting for help to arrive or how much daylight you have left depends on whether or not you are equipped to handle these threats.

Figure Out If You Will Move or Stay

In order to help you decide if you want to stay where you are or move on, you have to think about everything you know about where you are. Do you have a true sense of where you are in relation to where you should be? Do you think you might have just made a wrong turn a while back and are only a little bit off your course? If you are totally confident about where you were at about a mile back, turn back and go to that point and begin from there. If you feel lost, and you think you have been going in the wrong direction, it might be a good idea to just wait. Wandering around blindly might take you to a place where your rescuers have already looked, and this will take you even longer to get rescued. If you don't have any idea where you are, but you did tell your friend where you are going, it would make more sense to just stay where you are. You know that somebody will come looking for you.

If you think you know where you are at, but it is getting dark, but you don't have a flashlight, think about creating a shelter and begin walking when the sun comes up.

The decision of whether or not to move all depends on how knowledgeable you are about the area. If you know for certain that your car is parked on the road that runs north to south and you know that you have been heading east, you might be able to find your way back to your car. If you know a way to get to the area lake and you see a landmark you recognize in the distance, let those features guide you.

Be Ready for Anything

Any risk that you might face when you are in the wild can be reduced by being prepared, even if you don't think there is a chance of being trapped or lost. Today, we rely on battery-powered navigation and technology to help us, but there won't be a good signal in the woods. Since batteries die and cell phone signals are either non-existent or spotty, this is the last thing that you want to rely on when you're are in the woods with things that take batteries.

Begin by getting some knowledge about the region you will be in. Looking at maps could help you familiarize yourself with this area. You need to make sure you do this before you take off. When you know the place you are going better, including all landmarks and the terrain, you will not get turned around as quickly.

There are some things that you should take with you before you head out. This includes spare clothes, extra water, extra food, emergency shelter, a fire starter, a knife, water purification tablets, first aid kit, sun protection, headlamp, compass, and a map.

The extra water and food will make sure you don't go hungry if you spend a night in the woods, spare clothes are going to protect you from unexpected weather that might pop up or temperatures you weren't expecting. An emergency blanket can be used as a shelter. Headlamps are better than flashlights since they let you use your hand while building a shelter or starting a fire when it is dark. Any source of light is going to help you if you are stuck outside all night.

A whistle would be a good thing to take with you if you get lost. Its sound will travel a lot farther than your voice, and it doesn't use as much energy than screaming for help. Three blasts on the whistle is a universal signal for distress.

The most important thing that you could do before any trip is to share your plans with your family and friends. By doing this, your rescuers are going to know where they need to focus their search. Knowing that there is somebody who will miss you if you don't check-in at the right time could go a long way to reduce your panic.

CHAPTER 2

Where To Forage?

For people who live in urban communities, their local farmer's market is a great place for them to find and or sell wild foods. I love seeing wild edibles in neat little bundles on tables at the market. Paying five dollars for a bunch of ramps or $25 for a quarter of a pound of mugwort that you could find for free isn't appealing at all. How funny would it be if you get home with your bundles just to see the same thing growing just ten feet from your front door? This means you need to start looking around your neighborhood. As you learn new plants, you will start noticing it growing everywhere. Even if you live in the city, there might be places where you can find amaranth and lamb's quarters growing abundantly. You might find pounds of ripe black cherries and juneberries that you can get easily. Try to make friends with your neighbors and local community growers.

Talk with the local park caretakers and keep them from killing all the invasive edibles. You might be able to join a day of volunteering to weed your local park, and you get to take home what you dig up. If diplomacy doesn't work for you, I will go where I know it grows abundantly and forage what I want.

City planners and landscape architects are adding more edibles into their projects. If it isn't happening in your area, try to encourage them to do so. I am lucky enough to have a great local classroom, which is Brooklyn Bridge Park. Its designers and creators included many edible plants. You can't forage anything, but it is a great place to go and learn about edible plants. Plus, you will get the chance to teach others about them. You might be surprised to find some of these plants in your local nursery. This type of natural partnership needs to be more common in many areas.

If you have access to nature, you are very lucky. First of all, be respectful of nature. Forage in moderation and don't stomp that native plants. If you live in the city but want to forage for wild plants, talk with your local preserves and land trusts and make friends with some landowners who might let you forage on their land. They are out there; all you have to do is ask. If you own land, think about all the wild edibles that might be on your land and the best way to conserve and curate the land.

Even if a large part of your property is managed in ways like livestock, garden, or a lawn, there is a good chance that you have edible trees, shrubs, or weeds growing somewhere.

Some of the best forages for me have come from fencerows and farm meadows, but only do this if you have the landowner's permission. It is all about talking to the people who are in charge and asking them nicely. Normally, the answer will be an amused affirmative. Weeds in the middle of crop rows can be gathered as long as they haven't been sprayed with fertilizers or herbicides.

Growers and gardeners are very lucky. They have lots to draw on in terms of wild flavor. It all depends on your hardiness zone and microclimate, but most wild edibles can be cultivated. Bring some of your forages and plant them in your backyard.

State forests and parks are great places for you to explore. Even though state laws are going to vary, there is normally a fine for harvesting edible mushrooms and fruits for personal use. Picking mushrooms, nuts, and fruits is legal in US National forests. Harvesting fungi, plants, and other items are allowed in areas that are run by the Bureau of Land Management. Just check with your local office before you pick anything to make sure there aren't any preservation areas, protected species, or local statutes where you want to forage.

Harvesting fungi and vegetation in US National Parks are regulated. There are some parks that let you harvest specific nuts, berries, and fruits as long as you harvest them by hand, and they are used for your personal use. They could set limits to the quantity or size that is harvest. They could define a certain area where edible can be harvested and restrict the consumption and possession of wild edibles to an area of the park.

Private properties like farms, hunting grounds, and homes give you excellent foraging opportunities. You just need to make sure you ask the property owner's permission before you begin digging. It's a good idea to harvest near farms that use organic produce so that you don't ingest any dangerous chemicals.

Disturbing vegetation, fungi, and wildlife aren't allowed in nature preserves or centers. Just contact your local one to make sure.

Don't Harm Yourself

Foraging is something most people do on their own, but it does carry some big responsibilities. You have to make sure you inform yourself. Treat any and all information with an open mind and circumspection. Double-check, cross-reference, and remember that anybody will put things on the internet whether or not they actually know what they are talking about. Sometimes, published authors will get their facts wrong.

You have to be 100 percent sure you are identifying the mushroom or plant correctly before you eat it. Eat only the part of the plant that you are supposed to eat and during the correct growth stage. If the recipe you find says you should cook it, you need to cook it.

If you eat a new food you haven't eaten before, there will be a possibility you might have an allergic reaction. Try some of it, and then wait 24 hours to see if you experience any sort of reactions like vomiting or swelling.

Foraging can be done alone or with a group. Add wild foods into your menu means you have to find them, and finding them might lead to meeting with other people who have similar passions. If you are a loner who likes to do things along, then you will have a feast for a table of one.

Contamination

Even though wilderness and rural areas are thought of as pristine, looks can be very deceiving. Conventional farms, mining sites, power lines, roadsides, and railroad tracks are sources of contamination. Even areas that aren't close to human developments could be home to parasites and bacteria that is spread through animal droppings.

- Water Contamination

Foragers need to be cautious in areas that might contain animal feces. If you are harvesting plants in water, take some extra steps to prevent you from contracting giardia. This is an intestinal parasite that can cause watery diarrhea, vomiting, and cramps. Getting giardia is higher in areas where there is a lot of livestock since runoff from manure can get into the water system and can introduce the parasite. Wild animals have been known to carry this disease, too. This is how it got its nickname of "beaver fever." The good news is that giardia can be killed by cooking the foods, so it is fairly easy to stay away from that one.

- Pesticides

Herbicides are normally sprayed along power lines and railroad tracks to get rid of the overgrowth of vegetation. Conventional farms will spray pesticides on their crops to keep bugs off their crops. Some people think that ingesting edible wild plants that grow close to where pesticides have been applied is the same as eating produce from the grocery store. This is so false. The EPA is in control of the types of pesticides that can be used on food crops when they can be used, and the potential for residue. When you are foraging near conventional farms, power lines, or railroad tracks, you have no idea when pesticides were applied last. This means how much residue on the plant could be a lot higher than what you normally find in the grocery store. If it isn't a farm that produces food for humans to eat, they are spraying pesticides that are not safe to be eaten.

Pesticides, even the ones that are thought of as being safe for foods, have some risks for humans like some short-term toxicity to some long-term problems like reproductive disorders and cancer. Be leery of water around these places as they could contain pesticides from runoff. Agriculture is the main polluter of streams and rivers and the third polluter of reservoirs, ponds, and lakes.

How to Find Wild Edibles in a City

If a forager looks for wild edibles, what exactly do urban foragers do? Well, they look for plants that are edible in the city. Urban foraging is a large movement for guerrilla gardening, urban homesteading, and sustainable living. New York City, Portland, and San Francisco are leading the way by creating urban foraging classes and communities to help urban forages learn what plants are safe to eat and which ones aren't. Most of these groups will have websites where their users can share the information they found and create maps to show where they found the food. The Portland group lets browsers search for food by type of food or location and lets them add newly found items.

This is different from the growing "freegan" movement. A freegan is a person who doesn't want to participate in the consumer economy and goes "dumpster diving" to find food and other items. There might be some urban foragers who are also freegans, but most foragers don't feel comfortable digging through dumpsters for perishables no matter how safe freegans say these foods are.

Urban Plants You Can Eat

All the edible plants around you might vary a lot from what is available in your best friend's area. The most popular and easiest plants to identify, grow in various places.

- Dollarweed or Pennywort

This weed is completely edible. The round, young leaves have a crisp, pleasant taste similar to a snow pea or celery. The good news is that the plant contains a compound that has been proven to reduce blood pressure and relax blood vessels. If you are fortunate enough to live in the south, then you probably have these plants in your backyard. Rather than killing them with poison, try eating them. Since they do love damp environments, you need to wash them well before eating them.

- Rumex or Dock

Rumex or Dock is another common weed. There are several varieties that exist, and its leaves are best in early spring. It is a member of the buckwheat family, and it is related to chard. The leaves are edible, and you don't have to cook them, but there are times when they might taste very bitter. You should only eat them raw in moderation since they do contain a high concentration of oxalic acid. Oxalic acid can bind up nutrients in your food, and when you consume large amounts, it might lead to calcium and other mineral deficiencies. It would be best to cook the leaves before you eat them. Cook them just like you would spinach. The seeds of dock root are also edible, but the seeds do have a lot of chaff to them.

- Smilax

This is an easy weed to identify since it is a vine that has both tendrils and thorns. Most people think this plant is invasive, and they don't realize it is edible. Any tender part of the vine that will easily snap off with your fingers can either be cooked or eaten raw. Its taste is crispy, and light and you should treat it like asparagus. Even though you can eat it raw, consuming too much uncooked could make your stomach sour os it would be best if you steamed it and used it as a side dish. An extraction of the roots of some varieties was the original root beer or sarsaparilla. The root is starchy but a great source of calories and nutrients.

- Aloe

Most people have aloe plants for burns, but you can use the plant to make your own aloe juice. Aloe juice is very healthy because it contains natural anti-viral, anti-fungal, and anti-bacterial agents. This is why it is so great on burns.

There's a Map for That

"If you really love your peaches and want to shake your trees," there is a map that can help you find one. This goes for berries, nuts, vegetables, and lots of other edible plants, too.

Foragers Ethan Welty and Caleb Philips created an interactive map that will show you over half a million locations across the world where vegetables and fruits are free. They called their project "Falling Fruit," and it shows all kinds of trees that hang over fences, line city streets, and in public parks from New Zealand to the United Kingdom.

The maps look similar to Google Maps, but the foraging locations are shown as dots. All you have to do is zoom in and click on any of the dots. It will pop up a box that has a description of the bush or tree that you will find at that location. The description normally includes information about the best season to harvest the produce, the quality of the produce, and how much fruit the plant will yield. It also provides a link to the particular species profile on the USDA's website. Plus, all the advice about accessing this spot.

Welty, who is a geographer and photographer from Boulder, CO, compiled the locations from municipal databases, urban gardening groups, and local foraging organizations. The map can be edited by the public.

Welty considers himself a data geek, and he felt there is power in putting everything on one map. It is like having a narrow lens for the world.

With many countries boasting about having foraging destinations, it was almost impossible for them to find all the spots. They had to rely on their contributor's honesty when it came to listing trees in locations that might have been off-limits such as fenced-in parks or private property. In most cases, the contributors told foragers to ask the property owner for permission before harvesting produce. This map has over 6,700 entries so far.

The two states that they created this map to create a community for beginning foragers. There is a lot of value when you pull a carrot out of the ground or an apple off of a tree.

Using their skills to help other realize there is a fruit tree at the end of their street where they can get fresh apples, this is a simple thing they did to reconnect people with the way food works and to get them away from thinking they can only get fresh produce from the supermarket.

The map doesn't limit the entries to vegetables and fruits. It also lists dumpsters that have excess food waste, public water wells, and beehives.

The creators say they hope their maps and all of its contributors will continue to grow so large that it influences land belonging to cities and management plans.

They want to make people understand that they can forage for food in cities. They also help create a food forest like the Beacon Food Forest, among others, that can be found across the country. They want people to rethink what cities need to look like.

CHAPTER 3

When To Forage?

Many would be surprised to find foraging can be done year yard no matter where you live. If you live in places where the weather stays warm all year, then you will always have a decent supply of produce to pick. But for those who have four different seasons, what you have to forage for will change throughout the year. You also have to think about the fact that you will need to store some of what you forage to help get you through the winter when there won't be a lot of food available.

Make sure when you are foraging that you stay away from areas that have been or currently are old sawmills, gas stations, and industrial sites. Roadside soil will often be contaminated by car exhaust and residue.

Springtime Foraging

Springtime is the best time to forage for wild greens. There are a lot of different flavors, and they will make you wonder why you have ever settled for the leaves and herbs available at the grocery store. After having a long winter full of meatier and heavier meals, the body is ready to crave refreshing spring greens. If you are shopping at the store, you will find spinach, parsley, Swiss chard, arugula, bok choy, and kale. They are tasty, but they aren't really in season as their foraged counterparts.

You have lady smock, sorrel, and wild garlic, as well as jack-by-the-hedge, plantain, and purslane. You can also enjoy elder, hogweed, nettles, dock leaves, and dandelion. All of this and more is prevalent during the spring months. These are typically easy to be found, and most people refer to them as weeds, but you'll know better the next time you see them.

Wild greens tend to have a stronger flavor than greens you by in the store. They also have higher nutrient content, especially when it comes to magnesium, iron, calcium, folic acid, vitamin C, and beta-carotene. In rural areas, these wild greens make up most of the diet of those who live there. Most of these greens can be dried and saved for later in the season as well.

Let's take a look at some of the most prevalent greens to harvest during the spring and where they are commonly found.

- Nettle

Nettles are best when they picked in the spring as they are still young and delicate. They grow at river banks, in the woodlands, hedgerows, and in fields. Make sure that you have gloves when you pick nettle, though, as they do sting. In order to remove the sting, you will need to wash the leaves and then pound them in a pestle, but that is only if you want to eat them raw. Cooking them will also get rid of the sting.

- Wild Garlic

March to June is the best time to find wild garlic. It likes to grow in the woodlands and easily spotted through its pungent smell and long, broad leaves. You can eat any part of the plant, but most people want it for its leaves.

- Dandelion

This plant, which most people view as a weed when it's in their yard, is bitter. They are easily spotted from their bright yellow flowers. They grow abundantly pretty much anywhere. All parts of the plant, including the roots, can be eaten.

- Wild Sorrel

Sorrel can be found in most people's backyard. If not, they are common in woodlands, meadows, and grasslands.

- Wild Fennel

This is more commonly found in coastal areas and is easily spotted by its long stem and feathery fronds. They also have small yellow flowers that show up between April and July.

- Wild Chervil

Also known as cow parsley, wild chervil is a roadside plant. It can also be found in meadows and hedgerows. It looks very similar to hemlock, which is poisonous, so be cautious. This should be picked in winter, and early spring as the leaves are sweeter.

- Sea Purslane

This is a salt-marsh plant and is very salty. It grows near coastlines. It is great eaten raw and as a seasoning.

- Plantain

This is a plant that you can't mistake for anything else, and while you can enjoy it throughout the summer, it is more tender when picked during April and May. Just like a dandelion, it is full of vitamins K, A, and C, as well as iron and calcium. Broadleaf plantain is useful and can easily be spotted by its larger, low-growing leaves. You can dry the leaves to have them handy during the winter for some tea.

- Raspberry leaves

These are best when harvested during the spring before the flowers show up. Pinch only a few leaves from each plant, leaving the rest of them to turn into berries. This is a great plant for women because it helps with menstrual discomfort.

- Chickweed

These are one of the first greens to appear as winter makes its transition to spring. In fact, if the winters are mild where you are, it may not die back during the fall. This is great for salads. Chickweed grows in damp and cool places and typically germinates in recently disturbed soil.

- Hairy Bittercress

This plant is similar to mustard greens and is best used in the same way. Bittercress often shows up in the same areas as chickweed but will show up a little later on.

- Clover

You probably walk all over this tasty treat. All species of clover are edible. White and red clover are the most common types, but you can enjoy what grows abundantly in your hard. The younger they are, the easier they are to enjoy.

Summertime Foraging

Summer is the time of the year when plants start attracting pollinators so that they can produce fruit and so that their life cycle can continue on. This is when we have pretty flowers that can also be eaten. A lot of the plants that you harvested in the Spring may still be available at this point, but they are often a little more bitter or tough. But there are other delicious plants that you can enjoy during the summer. Be mindful of bugs and wash your plants very well.

- Lemon Balm

This plant is in the mint family, so it is easy to spot for its mint-like appearance. It has oval leaves that are hairy with a toothed edge. When crushed, they will smell distinctly like lemon. They will form white/yellow to pink flowers later in the season that bees love. The leaves on the top tend to be better for eating raw. You can take the older leaves, but make sure they are cooked or mixed with other items to mask their harshness.

- Borage

This is a hairy-spiny leaved plant that tastes a lot like cucumber. It produces five-petaled blue flowers, which you can pick by griping the black stamen to pull off the flower from the calyx. This is commonly found on waste ground and near gardens and hedges.

- Red Clover

Again, this can be found in any grassy area. When left to grow, it can reach nearly two feet tall. It flowers from May to September. It is very sweet, and the flowers should be picked when they are about a quarter of an inch in size.

- Oregano

This plant likes to grow in dry grassland, hedges, and woodland edges. It has hairy, green, oval leaves. It will flower, but it tastes better when you can harvest it before it flowers.

- Rose

All rose petals are edible but stay away from any plants that have been sprayed. Wild rose is often found in hedges and contains curved thorns. They often bloom from June to July. The field rose is similar, but is smaller and will have white flowers.

Autumn Foraging

Just like with a garden, nature will put out its biggest bounty during the fall. While greens are abundant during the spring, fall foragers can find fungi, nuts, roots, and fruits. Let's take a look at the tasty treats the fall has to offer.

- American Persimmon

Everybody is familiar with the Asian persimmon. What most people don't know is there is an American cousin that hides out in the hardwood forests of the eastern US. They are smaller but have a great taste. Make sure they are completely soft before eating.

- Pawpaw

These are the largest native fruits in North America. They are similar to cherimoya but are most commonly found in the bottomlands in the Eastern US. They are most commonly found on riverbanks in dense thickets. Pick once the skin turns yellow and the flesh is soft.

- Madrone Berries

You can find fruits in the west as well. Madrone berries are marble-sized fruits that come from iconic western trees. Make sure they are a deep red color. They are kind of dry, so they work better mixed with other things.

- Burdock

This is sometimes referred to as gobo. This is commonly grown in Asia, but it tends to be a weed in a pasture than a cultivated plant. It grows in every corner of America except for the Deep South. It has fuzzy leaves that are shaped like an elephant eat. Its taproot can be several feet in length. It is a biennial, so it only produces leaves the first year, then it will flower and set seeds during the second year. To get the best tasting burdock, harvest it at the end of its first year.

- Groundnut

This is a perennial leguminous vine that has an egg-sized tuber that you can use just like a potato. It prefers low-lying wetlands. Its vines climb over trees and shrubs in thickets from Texas to Maine and Florida to North Dakota. Harvesting their roots kills the plant, so be nice and replant a few of the tubers in the same place.

- Acorns

There is a lot of different oak trees through North America, and all of them produce edible acorns. But they are not a nut that you can just pop into your mouth off the tree. They are inedible until they have been processed a bit. To do this, place the acorns in a jar of water for a week or more, changing the water every day.

- Hazelnut

The hazelnuts bought in the store come from a cultivated European tree. But North America has its own native hazelnut, which is tasty but hasn't been developed into an agricultural crop. They will often grow on large shrubs that are found at the edge of forests and can be found in nearly every state except for the Southwest.

- Pine Nuts

These are unique to the southwest and California, where they are harvested from different pinyon pines. Pinyon pines are found at elevations over 5000 feet and should be harvested when the cones turn green to brown.

- Chanterelles

The season for wild mushrooms varies depending on where you live, but chanterelles are commonly found throughout American during the fall. They prefer cool weather and often emerge in the forest after a very good rain. They are one of the easiest mushrooms to spot. They contain ridges on the underside instead of gills like other species.

- Hen of the Woods

These are also called maitakes. You can find them throughout the states and are commonly located at the base of hardwood trees. They especially like oaks. They parasitize the tree, so they are more likely to be on dying or dead trees. They grow in huge layered clumps that are about the size of a chicken.

- Jerusalem Artichoke

Sometimes called a sunchoke, is a wild sunflower that is found in the central US. They tend to grow up to 12 feet tall. Their leaves are up to three inches wide and eight inches long, and their yellow flowers emerge around August and September. They grow nearly anywhere, but the best tuber production requires well-drained soil.

Winter Foraging

Most people think foraging will end once the weather turns cold, the leaves are gone, and most growth has come to a standstill. But what is a person supposed to do during the winter? When we are talking about winter, we are talking about the cool night in the desert or brisk winds you get in the southeast. Places that never see snow and rarely, if ever, see ice, don't really have a lot to worry about during the winter, but those who live from around North Carolina and up on the East coast, the Midwest, and the upper half of the West coast has less to work with.

I'm here to tell you, though, it is possible to find food during the winter, but let's take a look at some things that will complicate your winter foraging.

1. It's Cold

This won't only affect what you are trying to gather, but it is going to eventually affect you. The cold will also cause the ground to freeze, which is going to limit your access to tubers and roots.

2. There Could Be Snow

Snow can cover and obscure the things that you are looking for. You will have to know how to look for clues above the snow. An oak tree is a good indication that you could find acorns under the snow. Some oak trees that will hold onto a few of their leaves over the winter, so that will help.

3. It's Wet

Many people like harvesting cattails during the winter, but having to slosh through a foot or two of water and sticking your hand down into the water and mud is going to get old quick.

4. Less than 10% of Items are Available

If you are in an extreme winter climate, most things are going to be dead or not growing. Options will be limited to around 10%, depending on your location. In winter, we will often lose indicators that help us to discover food, especially leaves. However, there are still some indicators.

You need to make sure that you pay attention to the appearance and shape of the bark on trees, especially nut-bearing ones like black walnut, horse chestnut, and oak. Take the time to learn different barks and the characteristics of nut-bearing trees. A good clue would be if there is a squirrel's nest in the tree.

There are some plants that can continue photosynthesis under the snow. Scraping away the snow can reveal chickweed, wild onion, and dandelion. You can head into the water, but be careful. If you live close to the ocean, tide-pools during low-tide can help you find seaweed, kelp, and even shellfish if you eat fish. Ponds, freshwater springs, and creeks can have cattails, and crayfish and mussels if you want those as well. Make sure you are well dressed for water foraging during the winter. Remember, it's cold. You need to make sure that you dress warmly and in layers. You are going to experience varying degrees of rest and exertion, and you will need to be able to manage your perspiration.

The best foods to look for in the winter include:

- Cattails

The roots of these plants can be washed, peeled, and used similar to potatoes. You can also dry them and turn them into flower. If you decide to go looking for cattails in the winter, make sure it is the only thing you are planning on doing. Make sure you have some waterproof boots, insulated hip-waders, and some heavy-duty rubber gloves that go all the way up your arms.

- Horse Chestnuts, Acorns, and Black Walnuts

These will be on the ground at this point. Make sure that you soak them for about three days, changing the water every day, and then you can roast them, boil them, or dry them and turn them into flour.

- Rose Hips

These are typically bright red and around a quarter to a half-inch in diameter. You can make them into jelly or use them in tea.

- Mushroom

Different fungi may appear during the winter months, likely after a brief thaw. Try looking for them on rotting deadfalls. Make sure you don't pick toxic mushrooms.

- Wild Greens

While not as prevalent, there are some greens that will show up under the snow or poking through the leaves. Rinse and enjoy them.

- Wild Fruit like Crabapples and Plums

These are easy to spot as they will still be hanging on the tree. They are great as a jelly or as a juice.

Preserving Your Food

While you can find some plants to eat during the winter, the best way to make sure you have food for the winter is to act like a squirrel. Make sure that you forage enough during the year and preserve it so that you have it on hand when winter comes. Wild foods will last longer than store-bought foods. Most greens will last you about a month when kept in the fridge. Burdock flowers can last for two months, and the roots can last three to four months. There are different ways to preserve your food. Let's take a look at a few different methods.

- Blanch and Freeze

There are some wild edibles that will have to be blanched before you freeze them, so it is best to research each one individually. Others you can simply freeze as they are. Freezing is the easiest way to preserve foods. Rinse your plants in some cold water, shake off the excess, and chop them up. Add the chopped plants into an ice cube tray and then place them in the freezer. After they are frozen, place the plant-cubes into an air-tight container or freezer bag.

Another way to freeze the plants is to spread the loosely across a baking sheet and then freeze them. Once frozen, place the edibles in a freezer bag, seal, and keep it in the freezer. Now, these frozen treats won't be able to be used in salads because once they thaw, they will lose their integrity. However, they can be used in different cooking methods. Make sure you don't refreeze anything you have thawed.

- Drying

Drying is another popular method of preserving foods. If your wild edibles are clean, make sure you don't wet them. If they are not clean, briefly rinse the dirt and dust off of the foliage, shake any excess water off, and get rid of any damaged or dead foliage.

Take the stems of the plants and tie them together in a small bundle with an elastic band or string and then hang them upside down in an airy, dry place. Make sure they are not placed in direct sunlight. You want the bundles to be small and loose to allow for good air circulation. Elastic bands are best so that you can tighten up the band as the stems start to dry out and shrink. You can use paperclips to hang the bundles onto a rope or string.

The UV rays from the sun and moisture like frost and dew can end up discoloring the plants and reduce the quality of the herbs. That's what it is best if your dry your plants indoors in a closet, attic, or a small unused space. It can actually add a nice scent to a room.

If you don't want to hang your plants to dry them, you can spread them out across a clean window screen or some other type of screen. Place the screen between the backs of two chairs so that it can have has much airflow as possible. Flip the leaves over often to make sure everything dries evenly.

You can also use a conventional oven to dry your edibles. Spread the plants out into a single layer on a baking sheet and place them in the oven at the lowest temperature. Food dehydrators can do a great job, as well. Your plants are sufficiently dry when they crumble easily. After they are dry, you can separate the leaves from the stems. Keep the dried plants in mason jars with a tight lid. Keep them stored in a cool, dry place away from heat, moisture, and sunlight. When stored properly, they typically last one to two years.

CHAPTER 4

Tools For Foraging

All types of crafters have a cache of specialty tools, and foraging is no exception. The good news is, if you have a garden, flower garden, or yard of any type, you will likely already have some of these tools. Tools will make your life easier when it comes to foraging. Not all plants are made the same, so you will need slightly different tools for different types of plants. Sometimes plants just want to fight you when you try to harvest it, especially if you are trying to dig up roots during the winter. Tools make this process simpler, and it doesn't put as much stress and strain on your hands. Let's take a look at the types of tools you will want to have before you start foraging.

- Pruners

Pruners will be what you use the most often when you are gathering and processing foraged herbs. They are able to snip right through the herbaceous stem. They can also cut through roots, small branches, and twigs. You will find that you use these more than almost any other tool. If you are only able to buy one tool to get you started, pruners are the one tool that you should get.

There are different types of pruners out there. Look for high-quality pruners and make sure that their blades can be sharpened. Keep in mind the spring and blades can wear out. There is a brand called Felco that a lot of foragers rave about. They are long-lasting, and they even offer replacement springs and blades. Make sure that you keep them good and sharp because dull pruners can damage the plant.

Look for pruners that can help to reduce hand strain and fatigue. When your pruner handles are fully opened, they should not exceed the width of the of your grasp.

- Weeding Knife, Hori-Hori, or Japanese Garden Knife

This tool is a compact and heavy-duty, wildcrafting tool and is great for weeding. A hori-hori can be used to break up the soil and dig out small to medium-sized roots. The garden "knife" can cut through the majority of clay soils, and they can even be used to pry rocks up out of the ground. They can also be used to transplant and divide roots.

A good quality garden knife can last you for many years. Along with the pruner, you can purchase a holster for your knife so that you can keep it on your belt or person for easy access. The wooden-handled variety is believed to be stronger than the plastic ones, but if you tend to lose things easily, think about purchasing one that has an orange plastic handle to reduce your chance of losing it.

- Digging Fork

Digging fork, also known as a pitchfork, is the tool you will use to help you dig up most roots. The tines on the fork will loosen up the soil and help to life branching roots out of the Earth. A digging fork is a lot less damaging to roots than a spade or a shovel. The digging fork can be used in the garden to weed or loosen up the soil. Digging forks will have sturdy and square tines, unlike that of the hay or manure forks that have bendable, flat tines. You can find affordable digging forks at almost any big box store, but you do get what you pay for. The last thing you want to happen is to have your handle break after your second foraging session.

- Shovel

There is a good chance you already have a shovel hanging around your home somewhere. Having a few different types of shovels can be helpful. You will want to make sure that you have a least one long-handled shovel that has a pointed blade. Shovels will mainly be used to excavate large, tap-rooted plants like burdock. They are also helpful in heavily compacted soils.

- Compact Shovel

You may not want to lug around a full-size shovel with you, so a compact shovel is the way to go. There are a lot of underground edibles, such as burdock root, wild potato, cattail, and leeks. You can get compact shovels in various shovels. That said, if you are planning on doing a lot of foraging or are foraging bigger items, a full-sized shovel will make the work a lot easier.

- Kitchen Scissors

Having a sharp pair of kitchen scissors is a great tool to have for gathering tender-stemmed greens like cleavers, violet, or chickweed. Pruners can end up making a mess of the job as they are supposed to be used on tougher stems, and their blade reach is pretty limited.

- Pruning Knife

A pruning knife is a hook-shaped knife that will make short work of cutting stalks and vines. You can use straight edges, but a pruning knife will make things easier. They are purposely built with the hook shape to make slicing with a single stroke possible. This knife will allow you to cut low to the ground and in various terrains. It can also help protect your dedicated survival knife. If you can't find a pruning knife, a knife meant for cutting linoleum can do in a pinch.

- Pruning Saw

Having a foldable pruning saw can be a handy cutting tool for small to medium-sized tree branches and limbs. This is a good tool to have if you plan on gathering medicinal tree barks like black birch and wild cherry.

- Sharp Compact Knife

Having a simple sharp knife is great for peeling the bark off of medicinal trees. You will want to have a good quality folding knife or a compact knife that has a sheath so that you don't accidentally cut yourself.

- Assorted Baskets

You have to have something to hold your foraged foods, right? Baskets can help you out in many different ways. They are handy when you are gathering up and drying herbs, and they are pretty as well. It is helpful if you have an assortment of baskets. You can easily find baskets at thrift stores. Try to find some that have an open weave and are flattish and broad to add in ventilation for drying loose herbs.

- 5-Gallon Buckets or Tubtrugs

Buckets get used more often than you may think. They can be pulled out for large-scale harvests, such as blueberries and elderberries, and for harvesting muddy roots. Adding some water at the bottom of the bucket will help to keep leaves and stems of herbs fresh during a car ride.

You can use repurposed food-grade buckets. Both three and five-gallon sizes are great. You can also ask for some empty buckets at the food prep section or the bakery counter at your local grocery store. Five-gallon buckets can also be bought at hardware stores. Tubtrugs are pliable buckets with handles and can be quite helpful for harvesting foods. They tend to be on the expensive side, but they will last a good while.

- Gloves

Foraging can be rough on your hands, and your fingertips will love you for keeping a pair of gloves handy for any prickly situations, such as stinging nettles or a berry bramble. Plants aren't able to run away, so they have grown their own defense mechanisms that will wreak havoc on skin. A single encounter with nettles and the stinging sensation that occurs when you touch will remind you to have gloves with you at all times. Thorns are also a fun thing to run your hand across. There is an old joke that tells you how to spot the difference between a blackberry and a raspberry thorn. All you do is grab hold of the stem, pull your hand down its length, and if your flesh is still attached, you have a raspberry bush. But, if the flesh in on the thorns and your hand is torn to bits, it's a blackberry. This "joke" illustrates a good point. Carry good gloves to handle plants is worth your while. Having two different types of gloves handy is a good idea. A thin, supple pair can be helpful for delicate tasks. A thicker leather pair is better for the moments when you will need more protection.

- Heavy Duty Chopping Knife

Heavy-duty knives are necessary when you are chopping through those tough roots.

- Breathable Bags

This is an alternative to a basket. A breathable bag is another way to store what you have gathered and will allow air to reach the plants so that they don't start turning brown and decomposing. You can find these breathable bags at just about any store.

- Sturdy Vegetable Brush

You will want to have a sturdy bristled brush to scrub the soil out of the crevices and cracks of any roots you harvest.

- Loupe or Hand Lens

This isn't a must-have, but having a jeweler's loupe at a 10 to 20 times magnification can be helpful. This will allow you to look at small botanical parts and identify them more properly. They have a higher magnification than a simple magnifying lens. There are some that have LED lights attached, which is great for viewing plants.

- Water bottle or thermos

This won't help you to gather anything, but it will make sure you don't get dehydrated. No matter where you are foraging for plants, if you plan on being outside for any length of time, you need to make sure you have something to drink. A thermos is good if you are foraging in cooler weather. Some hot coffee or hot chocolate can help warm you up if you start to feel cold.

- Field Guides

One of the best ways to learn about plants is to have first-hand knowledge about it. When you have a reliable source with you, it can help to answer any questions and get rid of any uncertainty. If you can't find a guide, which most of us can't, using multiple field guides is the best way to go. There are some field guides that use pictures, and some will only use line drawings. There are a lot of them that will make the perfect plant sample and feature that when, in the real world, what you will find will be missing flowers, have fewer leaves, and things that aren't as bold. Once you have a plant tentatively identified, verify what you think it is with at least three other sources. You will want to make sure the descriptions line up across various sources and authors.

Ethical Foraging

While this may not be a tool that you can hold in your hand, it is a tool that you can have in mind to make sure that you don't cause harm when you are foraging. Knowing that lots of people want to learn how to forage makes me happy. The fact that people are willing to spend more time outdoors, interacting with nature, is a good thing. The process of identifying, harvesting, and preparing wild edibles can bring you closer to your ecosystem.

But it is a practice that can be full of abuse. Lack of education and overzealous actions can be dangerous to the person, as well as devastating to the environment. However, if you can keep just a few things in mind and follow the best practices, foraging is a rewarding and fun experience that won't harm the environment or you.

1. Know Your Environment

While people will obviously know that they can't start foraging for prickly pears in the rainy Pacific Northwest, it is important that you understand and know what is available or what could be "at-risk" in your area. Grab a field guide from a local library, online, or bookstore, and study up on your native flora. Pay close attention to the characteristics of the plant, their growing conditions, and when they fruit or bloom. The USDA plant map is a great tool to help your research if a wild edible grows in your area.

You should also seek out foraging walks provided by professional foragers and herbalists and get a spot in their next event. Local fishermen, farmers, and hunters tend to be great resources for finding abundant plant matter.

2. Have a Foraging Plan

You should never head out into the woods without having a plan. You need to have in mind the things that you are looking for and where the best places would be for you to start looking for them, then stick to that. If you happen upon something that interests you, take a note of your location, take a sample or picture, and then consult your identification materials and field guides when you are able to.

When you do this, it will likely save you a lot of trouble and danger of misidentification while also making sure that you are ethically harvesting because that plant could be in danger.

3. Harvest from "Clean" Areas

Make sure you forage in litter, spray, and pollution-free spaces. Don't harvest things off of the roadside, in industrial areas, along property lines, or in city parks. All of these areas have the potential for contaminants and pollutants. It is also best to find light traveled and untouched areas for your foraging. This will help to ensure that the harvest is safe and clean.

4. Identify

This cannot be stressed enough. You have to learn how to accurately identify herbs and mushrooms. Don't rely on a single characteristic, such as a leaf or bloom, to ID the plant. You need to use three or more points of identification. Consider the color, bloom, leaf, fruit, stem, bark, fragrance, branches, life cycle, location, spore print, and/or soil condition.

5. Be Conservative

The Majority of ethical foragers recommend that you only harvest around a tenth to a third of a certain patch of what you see, and never from one patch. For example, if you pass by a small patch of mugwort and see that it is sparse, small, and the only spot of mugwort in the area, don't harvest from it.

You should also look at the life cycle of the plant. If you take all of the white blossoms off of the elder tree during the spring, then there won't be any berries come fall. You should only harvest what you actually need. Exercise restraint even if it is hard.

6. Leave It As Good As or Better Than When You Found It

There is nothing more frustrating than seeing your favorite spots ransacked, spoiled, and pillaged by less appreciative people. Get rid of all of your garbage, and think about having another bag with you so that you can clean up any litter or messes that you come across.

Don't drastically change the landscape for your own means. You should not chop down limbs or tress. Don't pull a tree across a stream to make a bridge. Don't start driving off-road. Don't disturb animal homes, like dens or nests. Just don't do anything inappropriate.

You should report unsafe conditions to the correct authorities.

7. Prepare and Inform

Get the equipment and clothing ready that you are going to need for your foraging. Make sure you always let another person know where you are getting ready to go, and how long you plan on being gone.

8. Check Out The Legalities in Your Area

To be an ethical forager, you also have to be a legal one. Check out any regulations and laws in your area about where it is legal to gather mushrooms and plants, and if you are going to need a permit. Make sure you never trespass. Make sure that you have permission before you start foraging on private property.

If you are certain about the legal areas to forage, this is where you may want to talk to a hunger friend as they probably know the answer to your question. They may also be able to put you in touch with somebody who will let you forage on their property. You should also be aware of the hunting season schedule and take safety measures during those times.

By keeping these things in mind, it will help you to have fun, while also being safe and legal. Restraint, proper planning, and education are key characteristics in the ethical forager. Foraging helps to fulfill some basic and primitive urges, provide you with exercise, and adds activity to a normally inactive time. It helps to bring you closer to nature and strengthens your appreciation for the world around you.

CHAPTER 5

Identifying Plants

During the late summer and early fall months, the harvest is at its peak. Driving through the country, you will see roadside produce stands along the side of the roads. You can find all sorts of yummy foods like pumpkins, apples, peaches, berries, etc. When you forage for food, it lets you feel all the joy of eating local foods without having to wait for the farmers to harvest the food for you.

Foraging, as you have probably realized, is a lot harder than just walking into the woods and picking plants to eat. You absolutely can't do this because there are too many plants out there that are poisonous and could kill you.

Don't let this fact scare you. There are many benefits to eating plants found in the wild. You just have to know how to eat wild plants safely. Once you know this, you might just save yourself some money on your grocery bill. If you have a dog that can sniff out truffles or find a patch of rare berries, you are in business.

I like to forage because I get to be outside, get my hands in the dirt, and it helps me clear my head. I love to eat the clean foods that are free of pesticides while learning about the plants that are around me. But my most favorite part of foraging is getting to eat the food I find.

I love to target invasive plants and eating them. Some of the more edible invasive plants include burdock, lamb's quarters, and garlic mustard. Since man caused the problem with invasive species, it is about time we undo what we started.

How to Identify Edible Plants

It isn't a secret that everyone loves using apps to identify plants. If you are eating plants, you need to be 100 percent sure that the plant is safe for you. PlantSnap is a great app that will help you learn how to identify plants, but let's go one step farther to keep everyone safe.

You can begin with some of the most familiar plants, such as blackberries. Even though there are hundreds of different species and they are extremely hard to tell them apart, they are all edible.

You can use your apps, but it would help you if you can gain some knowledge of some basic botany skills. If PlantSnap gives you a few suggestions about some edible plants where your life, you can narrow your search down from toxic to tasty by noticing:

- General Notes: Watch out for the plant's shape and height.
- Stalk: Some plants might have spines, spots, or other markings on the trunk or stalk.
- Flowers, Fruits, Seeds, and Cones: Pay more attention to the plant's flowers, fruits, seeds, and cones as these are their reproductive parts and are normally easy to identify the plant by.
- Arrangement of Leaves: This can get confusing very quickly, but try to notice how the leaves are arranged with as much detail as you possibly can.
- Shape of Leaves: Do the leaves look like needles, are scaly and flat, or are they actually leaves? What does their outline look like?
- Micro-Ecosystem: The area where you found the plant, is it shady or sunny? What kind of soil was it in? Does this area get lots of rain? Can you tell if the water drains away from or towards the plant? What is the altitude of where you are at? Are you in a wetland, savannah, forest, or someplace else?
- Global Location: You shouldn't waste your time looking at a plant that is native to China if you are located in England, well, unless it is an invasive species that was brought from China.

If you can learn these particular traits, it can help you identify the right plants to forage.

Know Before You Go

Before you even think about heading into the woods to find salad fixings and berries, you have to know some of the main principles for foraging. Here is a quick overview of some things you need to know:

- Find invasive species if you can.
- Know ways to prepare the food when you find it. Some plants are only safe to eat after they have been cooked. Some edible plants will have toxic parts to them.
- Know the laws in your area. You can't forage for food in National Parks or private property.
- Leave some food for wild animals. Don't be that jackass who takes every single blackberry off of the bush. There are wild animals that depend on these plants, too.
- Know all the lookalike plants. Make sure you know of all the plants that may resemble the plant you are looking for.
- Know the food you want to find. If you know a little bit about what you are looking for, getting a proper identification will be a lot easier.
- Know the ecosystem. Try your best not to put any stress on delicate ecosystems.

Now that you have a fairly good idea about how to identify plants and why you should forage, it is time to get out there and see what you can find.

In the Woods

You have realized you are lost in the woods, you are days away from civilization, and you don't have many snacks left that you packed. How did you stray off the path? You are going to need to find food very soon.

You aren't going to have to look too far. Just look around you, there is a nutritious buffet waiting to be uprooted, cracked, and plucked. You have all sorts of edible plants just waiting on you.

In spite of what most people believe, most of the plants found in the forests in North America are safe to eat. The hard part is finding the plants that are tasty and nutritious when not cooked.

Most plants in the environment will be edible. They might not taste too good or give you many calories. Basically, you are going to need to be picky if you want a very decent meal.

Here are some tips on how to find plants that will satisfy your hunger while tasting good and ways you can stay away from their toxic cousins:

What Not to Eat

Plants are very tricky. Most are edible, but one mouthful of a bad plant could be deadly. That might be a bit of an exaggeration, but there are plants that you have to stay away from all the time.

Try not to get fooled by plants that look like edibles. Many wild plants look exactly like Italian parsley, such as hemlock. This plant is what killed Socrates, the Greek philosopher.

You don't have to know one poisonous plant; you just need to know what you are eating. Basically, this is saying you should only eat a plant if you can easily recognize and know that it is safe.

The plants listed below are the most common plants found all over the world, and you might be able to find them in your backyard. Being able to identify wild plants need to be at the top of your survival skills list.

- Ground Nut

Groundnuts are a member of the pea family, and they help fix the nitrogen levels in the soil. They aren't on the most popular food list because they do have a two-year cycle. Groundnuts like moist sandy soil near the banks of a river. They spread rapidly and can be found all across the United States. Their green parts look a lot like wisteria. Groundnut leaves will be pinnate and have between five and seven leaflets with smooth edges and hairless. The flowers have a musky fragrance. Groundnuts have an edible tuber that is made up of about 20 percent protein. The tubers will be sweeter in the fall, but you can harvest them at any time of the year. Trace the fragile stem to the ground and dig down about two inches. Then gently pull to unearth the tubers. Because their skins are thin, you don't have to peel them. Never eat them raw as they can cause gas and are very sticky. Cut them into small pieces and steam for 20 minutes. Check for doneness as you would a potato. You can save the stock for another soup if you would like.

- Wild Onion

This is probably the easiest food to identify. It is going to look just like an onion. The tops will be thinner and curly. You need to double-check this food's identity by smelling it. Many wild onions will grow in forests all across the United States, and they are a good source of food. If it doesn't smell like an onion, you shouldn't eat it. Your nose is very handy for finding poisonous plants. Stay away from any plant that smells like almonds, as this could be a sign of cyanide.

- Wood Sorrel

Most people are familiar with this plant as it can be found everywhere on the Earth except at the poles. There are over 800 species of this perennial, and it can grow between six and eight inches high. It has three leaves on each stem and looks very similar to clover. The wood sorrel will have a tart flavor. I remember eating this as a child in my backyard.

- Pony Foot

This plant looks like a pony's foot. It grows in wet, swampy areas. It can be easily found in most lawns. It doesn't have a strong flavor and is great to add to salads.

- Dollar Weed

This is a very common plant that most people don't want in their yards. It has a fresh taste similar to a mix of celery and carrots and can be put into any stock. It is a member of the carrot family, and the leaves are what you will eat. The roots and stems are too hard. It has bright green leaves that are rounded, and the edges are a bit wavy. It will produce small, white flowers in July and August.

- False Hawksbeard

This plant has a crinkly, veiny, edged leaves that are slightly curled. The plant comes up in early spring. In Florida, it will grow in the shade during hotter months. It does look like a dandelion because its leaves grow in rosettes, and it has yellow flowers. Hawksbeard is different from dandelion because their stem will contain multiple stalks, and each stalk will have multiple blooms. The young leaves can be eaten as a salad green. Older leaves can be added into soups and stews as an herb. You can find this plant from Pennsylvanian down to Florida and then west to Texas.

- Bacopa

This plant can be found in any semi-wet place in the world. It is a good health food that can affect neural development and regeneration. This can help with retaining memories. It has thick, small, succulent leaves that creep along the ground at about six inches high. The leaves will be rough when you touch them, and they smell citrusy like a lemon or lime. You can add them to some hot water and have a refreshing tea.

- Tree Nuts

What plants do you need to look for? You are surrounded by leaves, but it will take a wheelbarrow full to make you feel full. If your main goal is just survival, you need to find the fattier, calorie-dense parts of the plant. Tree nuts are a great option, and they are readily available in most woodlands.

If you live east of the Great Plains, look for hickory nuts. These are considered as the most calorie-dense food found in the wild. These are produced by deciduous, tall hardwoods. The nuts are very hard to crack. They will have an outer and inner shell. It will be worth your effort if you can get into them. They do taste a lot like pecans, and the pecan you find at your local grocery store actually comes from a southern species of hickory. You don't have to soak or cook them before you eat them. Just make sure the nut is veiny, just like a pecan. Buckeyes look a lot like hickory nuts, but they are extremely poisonous. The meat of the buckeye will be rounded and smooth.

If you find yourself in the American southwest forests, try to find pine nuts from the pinyon pine. This is a scrubby evergreen in the desert. It is a great option when you are hungry. These nuts are found inside pine cones. They are easy to harvest and taste like buttered kernels. The pine nuts you find in the supermarket are imported. Chefs and indigenous Americans have been using pinyon nuts for many, many years.

Don't forget about acorns. They are edible and were a food source for the Native Americans. You do need to prepare them first. You will need to take a rock to get the nut out of the shell. If you don't have a pot, you can use a clean sock to submerge the nut's meat into some water for a few days. The water will get rid of the tannic acid inside the nut. Too much tannic acid can cause stomach problems. If you do have a pot and are soaking the nuts in the pot, you will need to change the water a couple of times.

- Berries

Don't forget about wild berries when you find yourself lost in the wild. Just be careful when looking for berries as some varieties could make you very sick. Stick with berries that you recognize like raspberries, blackberries or other berries that grow in clusters. Other fruits like elderberries can be easily recognized by their purply-black berries that are shaped like an umbrella. Make sure you stay away from any white berries, and all of these will be toxic. Watch out for fruits that look like cherries or blueberries, they may taste great, but they are deadly doppelgangers.

Identifying Poisonous Plants

The best way to identify poisonous plants is by becoming familiar with the kinds that grow in your area like stinging nettle, poison hemlock, poison sumac, poison oak, and poison ivy. Poisonous plants come in many forms, and one rule of thumb isn't going to do it. Poison ivy grows on vines while poison sumac and oak grow in shrubs. Poison hemlock looks like giant bundles of parsley.

The rule I remember as a child is "leaves of three, let it be," but this only applies to poison ivy and oak. Poison sumac will have clusters of between seven and 12 leaves. To make things worse, there are other plants that also have leaves of three like the box elder sapling. This makes it even harder to tell these plants apart. Stinging nettle leaves will be heart-shaped, fine-toothed, and tapers on their ends. Stinging nettles will be extremely hairy; in fact, the whole plant is covered in hairs. They are even on the stems and the underside of the leaves. Even though you can eat this plant, you need to harvest them wearing gloves. If you touch any of the hairs, a chemical gets released that makes you feel like you have a swarm of bees stinging you. Stinging nettle will have very bright pink or yellow flowers. Poison sumac and oak don't have flowers, but poison ivy has clusters of small yellow flowers.

Plants in Your Yard

You have to realize that many of those beautiful plants that adorn your yard could be very toxic to you. Some can cause skin discomfort if you touch them, and you definitely don't want to eat them.

Monkshood is a great example of this kind of plant. It has beautiful stalks of purply-blue flowers that look like little wings. It does look beautiful in your garden, but it is very toxic. Just touching this plant can get the toxins on your skin, which in turn could cause numbness, tingling, and if it seeps into the skin could cause damage to the heart.

You Touched a Poisonous Plant, Now What?

First of all, try to figure out what plant you could have touched; wash the area with cool water and soap. Pat dry the area never rub. Make sure you don't touch any other parts of your body until you have had time to wash the affected area.

If a rash develops that doesn't go away with proper treatment, it can affect your heart rate or breathing, or causes more than just a mild reaction; you need to see a doctor as soon as possible.

CHAPTER 6

List Of Edible Wild Plants

Edible plants can be found anywhere from sidewalk cracks to forests. There is a plethora of free food out there just waiting for you to harvest them as long as you know what you are looking for.

Below you will find some plants that grow wild and are edible:

Alfalfa

This plant is part of the pea family. It can be found in fields across the country. It is very nutritious and has many benefits, like treating drug and alcohol dependency. The young shoots and leaves can be eaten raw.

Asparagus

This plant can be found growing almost anywhere in the world where there are gravelly or sandy conditions. It is dense in nutrients, vitamin C, fiber, calcium, and potassium. It's a great source of B vitamins, phosphorus, iron, magnesium. It can be eaten just like the kind you find in the grocery stores. It can be eaten raw or cooked lightly. You can fry, bake, boil, steam, or sauté it.

Blue Vervain

This plant can be found in most countries but is very abundant in the US and Canada. It likes moist conditions with full or partial sun. The seeds can be eaten when roasted and ground into powder. It can be a bit bitter in flavor. Leaves can be dried and used as a tea or tossed into soups, stews, and salads. The flowers can be used in salads, too.

Broadleaf Plantain

This is a member of the plantain family, and it can be found all over the world. It is full of vitamins K, C, and A. The whole plant is edible, but the young leaves are the best tasting. You can use these in any way that you would spinach-like sandwiches and salads. Some people have been known to eat the shoot of seeds when it has finished flowering. The more mature leaves can be eaten raw, but they are stringy and bitter. If you use the bigger leaves raw, think about taking out the veins first.

Bull Thistle

Bull thistle can be found in various environments but likes areas that don't get disturbed. It grows best where the soil stays slightly moist, but it has been known to grow in both wet and dry soils. You can find it on roadsides, fence lines, waste places, edges of forests, and pastures. Its rosettes can be troublesome in gardens and lawns. You can eat this plant. The young leaves and flower stems can be eating in salads or sautéed. The taste is a bit bland. You HAVE to remove all the prickles from the plant before you eat it.

Cattail

During summer and early fall, you could see lots of fuzzy, brown cattails swaying in the breeze in swampy places all over Canada and the United States. There are many parts of this plant that are edible. Cattails can produce more starch per acre than yams or potatoes. Unlike yams and potatoes, you are able to eat more than just the roots. Various parts of this plant can produce edibles at different stages of their development. The roots can be grilled until tender and eating like an artichoke. If you harvest the brown fuzzy catkins during spring while they are still hidden in the leaves, you can eat them like corn. Boil until hot and serve with pepper, salt, and butter. The stalks and shoots can be harvested and cooked like asparagus. You could clean them and spread on some peanut butter.

Chickory

This is a member of the dandelion family but is a bushy plant with blue flowers. You can eat the leaves and flowers raw, but the roots have to be boiled. Just like dandelion, it can be found all over the world.

Cleavers

Cleavers normally grow around the edges of fields and hedgerows. They could also be found near gardens, woodlands, pastures, disturbed areas, waste areas, orchards, and near crops. This plant is native to western Asia and Europe but is have made its way through North Africa, South America, Central America, Mexico, United States, Canada, and Australia. The stems and leaves can be used as a leaf vegetable even though it is sticky so they won't blend well in salads. You can use them in sandwiches or sautéed. The fruits can be collected, dried, and roasted to be used as a substitute for coffee. The stems and leaves could be dried as a tea.

Coltsfoot

This plant can be found in disturbed, open places. It can be found all over North America. The flowers of this plant can be eaten. You can toss them into salads to give it a great aromatic flavor. You can fill a jar with the flowers and then add honey to make a cough remedy. You can use this infused honey to sweeten any herbal tea. Flowers can be dried and chopped up to put into fritters and pancakes. Young leaves could be added to stews and soups.

Common Sow Thistle

This plant grows pretty much anywhere. It loves being in gravel banks, meadows, fields, roadsides, and cracks in driveways. This plant was brought to North American from

Europe. The roots, flowers, and leaves are all edible. They are best when the plant is young because the older plant becomes bitter. The root can be harvested while young. Roast and grind the roots as a substitute for coffee.

Common Yarrow

This plant can grow in South America, Africa, Australia, Asia, Europe, and North America. The leaves can be eaten cooked or raw. They have a bitter flavor when added to salads. They are best when harvested young. The leaves of this plant can be used in beers. You shouldn't consume these in large quantities. You can make tea from the leaves and flowers.

Creeping Charlie

This plant is native to Europe and southern Asia but was brought to North America to be used medicinally. It quickly adapted to its new surroundings and can now be found everywhere except in deserts and the coldest places in Canada. It is a relative to mint and is hard to control since it easily roots from any node on its stem. It does have a pungent minty flavor and works well when used as a herb. It is best eaten when the leaves are young. You can cook older leaves like spinach. You can also dry the leaves to be used in teas.

Crimson Clover

This plant can be found in most countries and loves growing in cultivated areas, lawns, fields, and meadows. The flowers and seeds are edible. The seeds can be sprouted and then used in sandwiches and salads. You can dry the seeds and grind it into flour. The flower heads can be used fresh or dried and put into teas.

Curly Dock

Curly dock is a great plant to eat as a snack. You can find these in almost every yard or meadow. Even though the leaves can be a good source of vitamins B and A, you need to eat it moderately as it could cause urinary tract infections and kidney stones. The roots of these plants can grow as far as two feet deep, so make sure you don't break them off when digging them up. Once you have them out, you can eat the roots and leaves fresh. It can be dried for later use.

Daisy Fleabane

This plant grows all over North American and has even naturalized to central Europe. The leaves are the only part of this plant that you can eat. They are a bit like hair, so they will have a fuzzy texture when you eat them raw. You can cook them like you would any green. The extract of leaves contains caffeic acid, which has neuroprotective effects on neuronal cells.

Dandelion

Every part of this plant can be eaten. This includes the roots, leaves, and flowers. The dandelion leaves are normally between five and 25 centimeters long and possibly longer. The flowers will be yellow to orange in color. Dandelions can be found all over the world. You can either eat all parts of this plant raw or cooked. They are a good source of vitamins K, C, and A. It also contains folate, B vitamins, and vitamin E.

Downy Yellow Violet

This violet is native in almost every area of eastern Canada and the United States. It also grows in some temperate regions of Asia and Europe. The leaves and flowers are edible. The root used to be used in ancient times for medicinal purposes.

Fern Leaf Yarrow

This plant grows in most countries. Even though the leaves are bitter, they can be consumed cooked or raw. The younger leaves, when mixed in a salad are best. Yarrow leaves can be used in beer making. Even though it is very beneficial and nutritious for any diet, you shouldn't eat too much regularly. You can make a tea from the dried leaves and flowers.

Field Pennycress

This plant can be found all over the world. This plant is edible, but it isn't very tasty. It is from the mustard family and has a very pronounced flavor. The larger the plant, the stronger the flavor, and after they bloom, they can be rather spicy. Cooking will take some of the edges off. You can put the young greens in salads if you want a bit of a kick. If you love eating radishes, you will love this plant.

Forget Me Not

The flowers are the part of this plant that is edible. They can be eaten as a snack or tossed into a salad. You can decorate dessert and garnish your meals with these pretty flowers.

These flowers grow well in zones five through nine in the United States. They can be put into baked goods as well as candied.

Garlic Mustard

This plant can be found anywhere in North American. It is a very invasive plant. The leaves are best when harvested young, and the roots taste a lot like horseradish. The leaves will be a bit bitter as they get older. A second-year plant can be eaten from early spring until about the middle of spring before the shoots have had time to harden. The seeds are great in spicy foods. Using garlic mustard all year long is a great way to keep this plant from spreading. The parts of the plant from older plants do contain cyanide. These parts need to be cooked well before consuming them.

Harebell

This plant grows throughout the US and Canada, along with being in Britain. This plant likes grassy, dry places. The leaves are the edible part of this plant and are best when put into a salad. You could add them to dips, or smoothies.

Japanese Knotweed

This plant can be found across North America. Not every part of this plant is edible. The shoots are what is edible, but they need to be harvested before the stems get woody and hard. The best time to eat knotweed is the middle of April until May. These shoots can be consumed raw and have a sour taste that is similar to rhubarb. You should cook them similarly as you would rhubarb. Think about knotweed in a cocktail, in preserves, or a pie.

Meadowsweet

This plant is normally found by rivers, lakes, wet woods, marshes, swamps, etc. It grows throughout Europe and North American, along with some parts of Asia. The young leaves are edible but are mostly used to make teas and soups.

Milk Thistle

This plant grows in most places around the world. The young flowers, roots, leaves, and stalks are all edible. The roots can be eaten cooked or raw. If you decide to eat the leaves, make sure you remove all the thistles first. The leaves make a great substitute for spinach. The flower bud can be cooked. The stems are best when peeled and soaked to help reduce their bitterness. You can use milk thistle-like rhubarb or asparagus, or you can put it in salads. The seeds of the milk thistle can be roasted and used as a substitute for coffee.

Prickly Pear Cactus

This plant can be found in most desert areas. It is a very popular plant as it is full of carotenoids, antioxidants, and fiber. The edible parts of this plant are its fruit, stems, flowers, and leaves. You can eat the whole cactus, either grilled or boiled. It can also be made into jams and juices.

Purple Dead Nettle

This plant can be found all over the world. This is a very nutritious superfood. The leaves are very edible, with its purple tops being very sweet. Because leaves are a bit fuzzy, so they are better when used as a garnish or mixed with other ingredients in recipes instead of being the star of the show. You can use purple dead nettle along with dandelion greens, chickweeds, or other "weeds" to make a "wild pesto." You can use this green as any other leafy green or herb.

Red Clover

You can find this plant all over the world in meadows, roadsides, grassy areas, vacant lots, pastures, and fields. The leaves can be put into salads or used as a tea. The part that is used the most is the flowers. Red clovers are the tastiest of all the species of clovers even though you shouldn't eat too many as it can cause bloating.

Sunflower

This plant is edible from its roots to its seeds. You can find them all across the United States. You can make everything from teas to salads. The roots can be chopped raw, marinated, mashed with potatoes, steamed, shredded into a slaw, sliced thin and fried, and roasted. The sprouts can be used just like soybean or alfalfa sprouts. The stalks taste a lot like celery. They can be chopped and added to salads or eaten like celery with peanut butter or hummus. The leaves can be used in salads or cooked like spinach. The petals make pretty garnishes, but you can eat them in salads, too. They have a bittersweet flavor that can be used to complement your dishes. Let's not forget about the seeds. The seeds are ready to be harvested when the flowers have turned from green to yellow. These seeds can be eaten raw, or soak them in salted water overnight and roast them for a great snack we all know and love.

Wild Black Cherry

This plant is native to North America. Even though the fruit is edible and can be used when cooking and in beverages, the rest of the plant has amygdalin and could be toxic if consumed. Wild cherry syrup is obtained from the bark and can be used as cough medicine. Wines and jellies can be prepared from the fruit.

Wood Sorrel

This plant looks like a clover but is not in any way related to clover. It is a great thirst-quencher and a great snack. The flowers of the plant can vary from bright yellow to green. The leaves are a wonderful source of vitamin C. All parts of the plants are edible and have a slightly sour taste that can be compared to lemons. This plant can be found all across North America, where there are dirt and sunshine.

CHAPTER 7

List Of Medicinal Wild Plants

Not only can you find a lot of edible plants to cook within the wild, but you can also find quite a few plants that can help heal minor ailments. All of these plants are also edible, but they are also well-known for their healing qualities. While these plants are considered medicinal, always consult with your doctor if you are experiencing any major health problems.

Bee Balm

This is an edible herb and is often used to add flavor to foods, and it makes an attractive garnish for salads. Herbalists use this plant as a stimulant, diuretic, diaphoretic, carminative, and antiseptic. It can be used to help with insomnia, menstrual pain, nausea, flatulence, sore throat, fever, gastric disorders, headaches, and colds.

Birch

The bark off of the birch tree, especially sweet birch bark, can be a great analgesic. The bark can be steeped to make a great tea. If you consume too much of this, it can end up causing nausea, upset stomach, or tinnitus. If you notice any of these things, you should stop taking it. It is recommended only to take 6 to 12 mg of salicin each day. Birch trees are found throughout most of North America and can be spotted by their white bark.

Black Cohosh

This plant is a member of the buttercup family and is native to North America. Black cohosh is used as a dietary supplement for menopausal symptoms and hot flashes along with premenstrual syndrome, cramps, and to help induce labor. It is thought to be used to help treat a range of other illnesses like high blood pressure and tinnitus. The flowers have a fairly strong odor are can repel insects. Large doses of black cohosh can cause poisoning and harm the liver.

Blackberry

Blackberries can be found pretty much anywhere. They are delicious in pies, but they can also help with diarrhea. They can also be made into a mouthwash because of their astringent nature. To make a tea, pour hot but not boiling water over some blackberry leaves and let steep for ten minutes. The normal ration is about two and a half ounces of leaves to one cup of water. You have to drink this when it was made. You can use this same tea as a mouthwash.

Black Walnut

The green hulls on the black walnut are commonly used in folk medicine. A tablespoon of the dried hull can be steeped in a cup of hot water to make a horrible tasting tea. Drinking a cup of this a day for a week can help you rid your body of parasites. The fresh hulls can also be used as a substitute for iodine tinctures on wounds and cuts. Black walnut trees are commonly found in the eastern US.

Burdock

The leaves and roots of this plant make a great liver tonic and can help to purify the blood and body. A lot of people have used burdock root to get rid of acne. It also has positive effects on many skin issues, such as eczema. You can make a tincture of the dried root in some alcohol and take ten to 20 drips every day. The fresh leaves and roots can be eaten after boiling them in water and getting rid of the water to get rid of the bitterness. You can find it around river banks and other disturbed habitats.

Chamomile

Chamomile is a commonly used flower for teas. It is so popular that you can easily find it on the tea aisle at the grocery store. It is a mild sedative and is great for treating insomnia and other nervous conditions. It is safe enough for teething children. It has anti-inflammatory properties and is great for arthritis. It can also help with menstrual cramps. Chamomile can also be used in shampoos and is common in slaves to help heal wounds and hemorrhoids. A compress of it can be used on skin inflammation to help with burns and sunburns. It is an annual herb that originally came from Europe, but can now be found on almost every continent. It likes sunny spaces and can be found from Southern Canada to Northern US and west to Minnesota. There are two types of chamomile, German and Roman. They both have the same healing properties, but Roman chamomile is perennial, whereas German chamomile is an annual.

Chickweed

This edible plant has been used as a folk medicine for hundreds of years for conditions like obesity, skin conditions, dyspepsia, inflammation, constipation, conjunctivitis, blood disorders, and asthma. The extract of this plant can be taken internally. It is normally used externally to treat sores and rashes. You can eat the
young shoots in salads. It can be found throughout North American and Europe.

Comfrey

The cooked and mashed roots of comfrey can be used as a topical treatment for sprains, burns, bruises, and arthritis. DON'T eat it. Research has shown that it can damage the liver if eaten. It contains pyrrolizidine alkaloids.

Echinacea

This is also known as coneflowers. It can be used as an herb and an ornamental plant. It grows best in USDA zones three through eight. Even though all of the plant is edible, the leaves and flowers buds are the parts that are most commonly used to make tea. Echinacea can help reduce inflammation, lowers blood pressure, helps calm anxiety, reduces the risk for breast cancer, help cells grow, controls blood sugar, can help fight the flu.

Elderberry

This tree can be found all over North America. The berries of this plant are very edible. They can be harvested and made into pies, syrup, jams, and wines. The whole flower cluster can be dipped in batter and fried. The petals can be eaten raw or made into a tea. The flowers can add an aromatic lightness and flavor to fritters or pancakes. The berries can be cooked down and made into a syrup. This syrup can help with cold discomforts and as a cough suppressant.

Evening Primrose

This plant can be found across North America. All parts of the plant are edible. It has a mild taste, but some say there is a rough aftertaste. The roots can be eaten cooked or raw, just like potatoes. If you soak the stem in water and boil it, the taste is similar to black salsify root. The leaves can be harvested from April to June before it begins flowering. You can eat them raw in salads or cooked in soup or like spinach. Native Americans make a tea from its leaves as a dietary aid. The oil from the seeds can be used to treat sore throats, digestive problems, hemorrhoids, bruises, acne, eczema, helps with PMS, eases breast pain before periods, reduces hot flashes, reduces high blood pressure, heart health, and reduces nerve pain.

Fireweed

This is a native plant found in the Northern Hemisphere. You can identify it easily by its erect, smooth, red stems. It has unique leaves that have a circular pattern on them that doesn't stop at the end. It has beautiful purple flowers. All those parts of the plant that grows above the ground can be used to make medicine. It can be used for swelling, pain, enlarged prostate, wounds, tumors, and fevers. It could be used as a tonic and astringent. The early shoots can be enjoyed raw or cooked lightly. Harvest the leaves when they are pointing upwards and close to the stem. These leaves can be pinched off and eaten like any leafy green. Once the plants get larger, the leaves get fibrous and unpleasant. The flower buds can be added to salads for a colorful addition.

Ginseng

Ginseng has a long history of being used for medicine that goes back more than 5,000 years. It can help the body adapt to emotional and mental stress, hunger, cold, heat, and fatigue. Ginseng is great for our metabolism, and it can promote well being. It also has a reputation for being an aphrodisiac. These facts seem to be based on the fact that it can relax someone who is very tense. If you have chronic back pain or TMJ, you can add this to a cup of slippery elm and catnip tea.

Henbit

This plant can be found throughout the US and Canada. It can also be found in Greenland, western Asia, South America, and Australia. You can eat this plant cook or raw, along with being used in teas. The leaves, flowers, and stems are edible, and even though this is part of the mint family, most people state that these plants taste a bit like raw kale. It is high in fiber, vitamins, and iron. This medicinal plant can be used as a stimulant, laxative, febrifuge, excitant, diaphoretic, and anti-rheumatic.

Herb Robert

This plant has a history of being a "miracle" plant, but it has been connected to witchcraft. You can eat this plant. The leaves can be eaten raw or steeped in teas. Due to its odor, you can plant a ring of Herb Robert around your garden to keep rabbits and deer away. You can rub the leaves on your skin to repel mosquitoes. It can support the immune system since it is an antioxidant. It can lower blood sugar and can help with diabetes. European herbalists have used it to treat toothaches, bruises, wounds, dysentery, gout, nosebleed, jaundice, conjunctivitis, and cancer. Like all other geraniums, it can help alleviate diarrhea. This herb can also be used to treat foot-and-mouth disease in animals.

Honeysuckle

This has been used as a medicine for thousands of years in Asia. A decoction of honeysuckle stems can be taken internally to treat hepatitis, mumps, and rheumatoid arthritis. The flowers and stems can be used together in an infusion to treat dysentery and URIs. The flowers bud can be turned into an infusion to treat various ailments like enteritis, colds, bacterial dysentery, tumors, and syphilitic skin diseases. Flower extracts can lower blood cholesterol. It can also reduce blood pressure. It can be used as a tuberculostatic, antiviral, and antibacterial. The glowers can be applied straight to sores, infectious rashes, and inflammations.

Jewelweed

If you ever find yourself in contact with poison sumac, ivy, or oak, look for some jewelweed. Crush up the purplish stalk into a paste and rub it over the affected skin. Leave it on for two minutes, and then wash it under clean water. If you can do this within 30 to 45 minutes after exposure is better. You shouldn't have much of a reaction. If it took you a little longer to find some jewelweed, you might still have some relief if you use it as a wash. Jewelweed can also be used to cool the itch on an already established breakout. It can be found in moist, semi-shady areas.

Lamb's Quarters

This plant can be found all around the world. It is full of essential minerals and vitamins, vitamin A, vitamin C, and a good source of B vitamins, including niacin, riboflavin, and thiamine. You can use this plant just like spinach. You can use them fresh in juices, salads, or any recipe that might call for greens. They are best when harvested young, but as the leaves mature, the flavor could change because of the greater potency of oxalic acids. Taste the leaves before you harvest a bunch to make sure their flavor is what you want it to be. You can chew the leaves and put it sunburn, inflammation, injuries, minor scrapes, and insect bites. A tea made from the leaves can help with loss of appetite, stomach aches, internal inflammation, and diarrhea. This tea can be used as a wash to help heal skin problems. The leaves eaten either raw or fresh can help support the blood system and heal anemia.

Lavender

Lavender is normally used as a fragrance these days, but it has been used since ancient times to repel insects. It has also been used to treat skin disorders, burns, and bug bites.

It can help relieve rashes and the itching that goes along with them. It can also reduce swelling. Crush some fresh leaves and apply them to the affected area. You could fill a jar with some dried leaves and cover them with some olive oil. Allow this to sit between six and eight weeks. You can then use the oil for any skin problems. You should never take lavender internally by small children, nursing, or pregnant women.

Lemon Balm

Lemon balm can be found all around the world. You can make the best-tasting lemonade you've ever had by adding some muddled lemon balm leaves to your lemonade. This plant is also good for cold sores and can be used to fight insomnia. Germany's version of the FDA has lemon balm listed as being a better remedy for cold sores than the leading prescription. Crush some fresh leaves and place them over the sores. You can use a cream that contains a very high concentration of lemon balm.

Lobelia

A word of caution, do not take too much lobelia as it can result in death and less-severe side effects. It has been used to help treat respiratory complaints, like asthma, pneumonia, whooping cough, and bronchitis. It is an expectorant, emetic, analgesic, stimulant, sedative, and antispasmodic. It is a perennial herb native to Eastern North America from Maine to South Dakota and south to Missouri and Texas. They bloom from July to November and can grow to three feet high.

Willow

The weeping willow tree is an easy to spot tree common throughout North America. While it is not native, it does well in moist areas. You can spot the tree by its droopy branches and twigs. The bark and leaves have been used to make medicine for centuries. Boiling a palmful of the leaves in a cup of water for ten minutes will make an astringent. Soak a cloth in it to apply to skin problems you are having when you have no other medical treatments available. Bark scraping can be soaked in a cup of hot water for ten minutes and then drunk to stop diarrhea. Take a sip of it every two hours, and continue this until the symptoms go away.

Yarrow

Crushed flowers and leaves can be placed on scratches and cuts to stop their bleeding, and it reduces the chance of getting infected. The leaves encourage clotting, and they are an antiseptic.

CHAPTER 8

List Of Poisonous Plants

Lastly, we will go over some of the most common poisonous plants. It is just as important to know what plants out there will hurt you. Not all plants have the same level of toxicity. In the list below, we will use the numbers 1 through 4 to represent the toxicity levels. One is major toxicity, meaning it can cause serious illness or death. Two is minor toxicity, which can cause minor side effects, like diarrhea or vomiting. Three is oxalates, which means its sap or juice contains oxalate crystals, which can irritate the throat, tongue, and mouth. Four is dermatitis, which means it simply irritates the skin. There won't be very many threes and fours.

African Boxwood

The leaves on this level two plant are the most dangerous. Touching the plant can also cause skin irritation. When eaten, you could experience convulsions, dizziness, diarrhea, vomiting, or nausea.

Angel's Trumpet

A level one plant and the entire plant should be avoided. When ingested, it can cause hallucinations, tachycardia, paralysis, memory loss, and ultimately death.

Azalea

A level one plant and the entire plant should be avoided. While small amounts won't cause much more than a stomach upset, if large amounts are consumed, or if you eat honey made from azalea pollen, it can cause serious problems. This normally happens in areas where there is a dense population of these plants. The honey is often called mad honey because of the confusion it will cause.

Bird-of-Paradise

This is considered a level two plant. The seeds of the flower contain toxic tannins, and the leaves have hydrocyanic acid. It is typically considered non-toxic towards humans and is more harmful to pets, but if large amounts are ingested, it can lead to dehydration or choking.

Bitter Nightshade

This is considered a level one plant, and the berries are the most dangerous. It is a woody perennial. It can reach six feet in height, and its purple flowers will turn into roundish berries. They contain solanine, a poison, which can end up causing headaches, stomachache, drowsiness, lowered temperature, trembling, vomiting, and diarrhea.

Black Henbane

A level one plant, but the seeds and leaves are the most dangerous parts. No part of the plant should be consumed. It can cause salivation, nausea, headache, diarrhea, vomiting, convulsions, rapid pulse, and coma.

Black Locust

A level one plant, and the seeds, leaves, and barks can cause the most problems. Fatal cases of black locust ingestion are rare, but recovery from its side effects tend to take several weeks.

Bushman's Poison

A level one plant and its sap is the most poisonous part. The sap of the plant was often used by the bushmen to create a cocktail that they would dip their arrows in to help them kill when hunting.

Castor Bean

A level one plant and its seeds, leaves, and bark are the most dangerous. This plant is what is made to create castor oil, which is an emetic. While it may not kill you to take castor oil, as long as you take the recommended dose, consuming the plant can, especially the bean.

Checkered Lily

This is a level one plant. The bulb of this plant is the most dangerous. The bulb contains poisonous alkaloids. In fact, all lilies and daffodils contain this dangerous alkaloid.

Chinaberry

This is a level one plant, and its fruit is the most dangerous. Symptoms of ingestion are diarrhea, vomiting, breathing difficulties, and paralysis. Birds and cattle are the only animals that can consume the berries without harm.

Chinese Lantern

This is a level one plant, and its fruit and leaves are the most dangerous. Sometimes referred to as the ground cherry, the pods of the Chinese lantern plant can be highly toxic and fatal.

Climbing Lily

This is a level one plant, and no part of the plant should be eaten. The tubers of the plant resemble yams but are one of the most toxic parts of the plant. It can be fatal.

Coral Bean

This is a level one plant, and no part of the plant should be eaten. Some people say you can eat certain parts of the plant, but it is safer to avoid it altogether. Some parts are hallucinogenic and narcotic.

Daphne

You should avoid the fruit and most parts of this level one plant. Chewing on any part of the berries, bark, foliage, or flowers can prove to be fatal.

Deadly Nightshade

This level one plant is also known as belladonna and is one of the most famous poisonous plants. It contains scopolamine and atropine in is roots, berries, leaves, and stems. The berries start green but turn black when ripe. The tricky part is they are sweet and juicy. A couple of berries can kill a child. An adult would have to eat somewhere between ten to 20 berries before they die, but they are going to be very sick even if it doesn't kill them.

Death Camas

You should avoid all parts of this level one plant. The mature leaves and bulbs are the most toxic. The plant contains several steroidal alkaloids. It can cause ataxia, tremors, muscular weakness, and prostration.

English Laurel

This level one plant's seeds should be avoided, but any part can be poisonous if enough is ingested. It causes potentially fatal respiratory problems.

English Yew

This level one plant should be avoided, but its seeds, leaves, and bark can cause the most problems. It is mainly used as an ornamental plant, but the ingestion of any part can be fatal.

European Mistletoe

No part of this level one plant should be consumed. Ingestion of the plant can result in death. American mistletoe, though it has a common name, is not as poisonous, and will likely only cause minor intestinal upset.

Flowering Tobacco

This is another level one plant. Its leaves are the most dangerous parts. Just like any other tobacco plant, it contains anabasine, nicotine, and other alkaloids. Its toxins are easily absorbed through the lungs and stomach. When too much is ingested, it can cause elevated heart rates, agitation, and coma.

Foxglove

This is a level one plant, and no parts of the plant should be eaten. While it has been used as a heart medicine, it is actually very poisonous. It can be safe in the right doses, but it is very easy to overdo it. Symptoms of poisoning are low blood pressure, collapse, and an irregular heartbeat. You could also experience drowsiness, intestinal upset, rash, lethargy, headache, depression, and blurred vision.

Heliotrope

The seeds of this plant are a level one. The plant can cause gastric distress. It also contains liver toxins, which can end up causing liver damage.

Holly

This is a level two plant, and the berries are the most poisonous parts. If swallowed, it can result in diarrhea, vomiting, drowsiness, and dehydration.

Poison Hemlock

No part of this level one plant should be consumed. This plant is very dangerous for the fact that it looks a lot like other non-poisonous plants. At first, it will cause salivation, nervousness, and tremors, but if it moves into a depressive phase, it can result in death.

Poison Ivy, Oak, and Sumac

These are considered a level 4 because they contain what is known as urushiol. This will cause an allergic reaction that can cause a rash on your skin. It is found every in the US, except for Hawaii and Alaska. They can be spotted by their leaves of three.

Japanese Pieris

No part of this level one plant should be consumed. This plant falls into the same category as azaleas and rhododendrons. If consumed, it can cause abdominal pain, vomiting, and excessive salivation after only six to eight hours after consumption.

Japanese Yew

This is a level one plant, and most parts of the plant should be avoided. This plant contains taxine A and B, which can prove to be fatal when ingested. The main symptoms are vomiting, breathing trouble, and tremors.

Jerusalem Cherry

The fruit of this level one plant is dangerous. This is a nightshade and is known as a pseudocapsicum. Its main poison is salanocapsine, which is a lot like other alkaloids found in this genus. It is not typically life-threatening. It can cause gastrointestinal upset and can cause vomiting.

Jimsonweed

No parts of this level one plant should be consumed. Poisoning most commonly happens when somebody eats or sucks the juice out of the seeds from the plant. It can also be toxic through touch, though not as common.

Lupine

The seeds of this level one plant can be dangerous. This plant is also known as bluebells. It can cause intestinal upset in humans. The poison is found in the foliage of the plant, but it is more concentrated in its seeds.

Monkshood

This is a level one plant and is considered poisonous to the touch. Most of the time, touching the plant won't cause any serious effects; it can cause tingling or numbness in the hand. If eaten, especially the seeds and roots can cause heart problems, tingling sensation of the skin, diarrhea, coma, and possibly death.

Myrtle

This is considered a level two plant. It is also known as common periwinkle. It causes systemic toxicity, which can cause mild abdominal discomfort to serious cardiac problems.

Nephthytis

This is a level three plant. This plant won't necessarily kill you if you eat it, but it can cause a lot of unpleasant symptoms that can occur for two weeks after eating it. It can cause difficulty swallowing, swelling of the throat, drooling, and burning sensation of the tongue, lips, throat, and mouth.

Pokeweed

All parts of this level two plant are toxic to humans and animals. The roots tend to be the most poisonous. The stems and leaves cause intermediate toxicity, and the berries are the least toxic. Children are most commonly poisoned because they will eat the berries. That said, some people will eat pokeweed, but only after the leaves have been properly prepared.

Tansy

This plant is considered a level four. It mainly only causes contact dermatitis for people who are allergic to the plant. If the flowers and leaves are consumed in large amount can cause poisoning because it contains thujone, which can cause brain and liver damage.

Water Hemlock

Not parts of the level one plant should be consumed. This is one of the most poisonous plants in North America. Only a very small amount of the plant can poison humans and livestock. Its main toxin is cicutoxin, which affects the central nervous system.

Woody Nightshade

This level one plant should be completely avoided. This is sometimes called bittersweet. Its egg-shaped red berries and foliage are poisonous. They contain solanine, which will cause convulsions and death if large doses are taken.

Wormwood

This is considered a level four plant. It is part of the Asteraceae family, like marigolds and ragweed, which means if you are allergic to those plants, you may experience an allergic response to wormwood. It is also toxic to the kidneys if too much is consumed.

CHAPTER 9

Recipes For Foraged Plants

<u>Stinging Nettle Spanakopita</u>

You Will Need:

- Melted butter, .5 c
- Phyllo sheets, 18
- Grated nutmeg, .25 tsp
- Chopped parsley, .33 c
- Beaten egg, 2

- Grated parmesan, .5 c
- Crumbled feta, 1.5 c
- Chopped ramps or scallions, .75 c
- Melted butter, 2 tbsp
- Fresh stinging nettle leaves, 8 c

You Will Do:

1. Start by steaming the nettle leaves until they are wilted. Remove them from the steamer and let them completely drain. Place them on a cutting board and chop.
2. Next, add the ramps or scallions with two tablespoons of melted butter in a skillet and sauté. Add in the nettle and sauté for a few more minutes.
3. Set the skillet off of the heat and add in the nutmeg, parsley, egg, parmesan, and feta and mix everything together.
4. Take a 9 by 13 pan and lightly coat it with some melted butter. Unroll the phyllo and cover it with a damp dishtowel. This keeps it from drying out. You will only be working with one sheet of phyllo at a time. Work as quickly as you can.
5. With a pastry brush, brush the butter on the first phyllo sheet. Place it onto the buttered pan, off-centered, so that it covers most of the bottom and comes up one of the sides.
6. Brush your next sheet with butter and lay it on the other side of the pan. Do this six more times, each one of them being placed so that they come up the other side of the pan.

7. Now the next four sheets, once brushed with butter, should be placed in the very center bottom of the pan.
8. Spread the nettle mixture evenly over the phyllo.
9. Now, work as quickly as you can to butter and place the reaming sheets of phyllo on top of your filling. Take the phyllo that is up the sides of the pan and fold them down on top of the phyllo you just places. Brush with butter.
10. Score the top of your phyllo so that it is cut into 12 pieces, but make sure you don't cut down through the filling. Allow this to bake for 35 to 45 minutes at 375, or until everything is golden.

Dandelion Fritters

You Will Need:

- Tallow, lard, or other fat for frying
- Egg
- Melted butter, 1 tbsp
- Milk, .5 c
- Salt, .5 tsp
- Baking powder, .25 tsp
- Flour, .5 c
- Dandelion flowers, 1.5 c

You Will Do:

1. Start by adding your chosen fat to a pot and heat it up so that it is ready for frying.
2. Combine the salt, baking powder, and flower. Mix in the egg, melted butter, and milk.
3. Working one at a time, coat a dandelion flower in the batter.
4. Drop the coated flower into the hot oil and fry until browned, turning them once.
5. Lay them out on some paper towels to soak up the excess oil.
6. These are great with maple syrup or honey.

Seaweed Salad

You Will Need:

- Toasted sesame seeds
- Sliced green onions, 3
- Sea beans, .3 lb
- Various seaweeds, 1 lb

Dressing:

- Sugar, 1 tbsp
- Soy sauce, 1 tbsp
- Sesame oil, 1 tbsp
- Rice wine vinegar, 2 tbsp

You Will Do:

1. Start by whisking all of the dressing ingredients together until the sugar is dissolved.
2. Boil a pot of water and add in the sea beans. Boil for a minute and then place them in a bowl of ice water. Boil the seaweed for 15 seconds and then place it into an ice bath. Dry the vegetables and place them into a bowl.
3. Add the rest of the salad ingredients into the bowl and pour the dressing over top.

Wild Mushroom Ragu

You Will Need:

- Grated hard cheese
- Pepper
- Oregano sprig
- Mushroom soaking liquid, 1 to 3 c
- Dried oregano, 1 tbsp
- Red wine, 2 c
- Tomato paste, 2 tbsp
- Chopped garlic, 4 cloves
- Finely chopped onion
- Diced bacon, .25 lb
- Olive oil, 3 tbsp
- Dried wild mushrooms, 2 to 3 oz

You Will Do:

1. Soak the mushrooms in a bowl of hot water. Cover them so that they stay submerged. They need to soak for at least 30 minutes, or up to a few hours depending on how big they are. Take the mushrooms out once they have been rehydrated and rinse. Place them on some paper towels and squeeze them dry.

2. Pour the soaking water through a sieve so that you get rid of the girt.

3. In a pot, add the butter and bacon. Cook until crisp. Take out the bacon and add in the onions. Cook until they start to brown. Add the mushrooms and garlic, stirring well. Cook for a few minutes until the garlic browns up. Mix the bacon back in along with the tomato paste.

4. Add in the oregano and red wine. Once boiling, let it cook down for five minutes. Add the mushroom liquid.

5. Taste to see if you need some salt. Let this simmer until the liquid has reduced a bit, and the soup is thickened up some. This will take about 20 minutes. Then, turn to low, cover, and cook for an hour.

6. Cook some ragu pasta, and once the mushrooms are done, top the pasta with the sauce. Add some grated cheese, pepper, and oregano.

Stir-Fried Dandelion Greens

You Will Need:

- Pepper and salt
- Olive oil, 1 tbsp
- Red pepper flakes, .25 tsp
- Minced garlic, 1 to 2 cloves
- Fresh, washed dandelion greens, 3 to 4 cups

You Will Do:

1. Add the olive oil to a skillet and heat. Add in the red pepper flakes and garlic, and stir-fry them so that the garlic doesn't brown. Once the garlic has softened, add in the greens.
2. Stir the greens so that they are well coated in the oil. Continue to stir them around for about five to eight minutes. The goal is to wilt them, but not let them become soggy.
3. Transfer to a dish and enjoy.

Purslane Tacos

You Will Need:

- Minced garlic clove
- Diced small jalapeno
- Diced tomato
- Purslane, 1 c
- Small onion
- Coconut oil, 1 tbsp
- Beaten eggs, 2
- Salsa
- Queso fresco
- Warmed corn tortillas
- Pepper
- Salt

You Will Do:

1. Beat the eggs and season them with some pepper and salt. Place to the side.
2. Heat a skillet and add in some oil. Cook the onion for a couple of minutes, or until it turns translucent. Add in the purslane. Stir and cook for two additional minutes.

Add in the garlic, jalapeno, and tomato. Cook everything together for a minute. This will help to cook the juice off of the tomato.

3. With a spatula, push the mixture to the side to make some space to cook your eggs. Scramble your eggs for a few minutes and then mix the eggs and the purslane together. Add some more pepper and salt if needed.

4. Serve the mixture in some corn tortillas and top with salsa and queso fresco.

Zucchini and Purslane Soup

You Will Need:

- Pepper and salt
- Thickener of choice, 2 tbsp – cornstarch or flour works well
- Heavy cream, 1 c
- Chicken broth, 4 c
- Shredded basil, 2 tbsp
- Zucchini, 3 lb
- Washed purslane leaves, 2 c
- Minced garlic, 3 cloves
- Diced medium onion
- Extra virgin olive oil, 2 tbsp

You Will Do:

1. Start by heating the oil in a pot and add in the garlic and onion. Cook for a couple of minutes until they become fragrant. Add in the sliced zucchini and sprinkle in a bit of pepper and salt.
2. Cook until the zucchini has become soft. Add in the purslane and basil. Cook for a few more minutes, stirring often. Make sure that they don't brown.
3. Once cooked, add the vegetables into a blender or food processor and puree.

4. Heat the broth up and add the pureed vegetables into the broth.

5. Combine your chosen thickener with the heavy cream and whisk it into the soup. Cook everything for three minutes, or until the soup has thickened.

6. Taste and adjust any flavors that you need to. Enjoy.

Purslane Salad

You Will Need:

- Pepper and salt
- Zest of a lemon
- Buttermilk dressing, 3 tbsp – use your favorite
- Blue cheese, 4 oz
- Fresh blueberries, 1 c
- Purslane leaves, 4 c

You Will Do:

1. In a salad bowl, add the purslane and drizzle with the buttermilk dressing. Toss to coat the purslane.
2. Fold in the blue cheese and blueberries.
3. Add the pepper, salt, and lemon zest. Stir everything together and enjoy.

Stinging Nettle Soup

You Will Need:

- Heavy cream or buttermilk, 1 c
- Peeled and quartered potatoes, 5 to 6 small
- Chicken stock, 4 c
- Salt, 2 tsp
- Dried thyme, .5 tsp
- Chopped onion, .5 c
- Chopped celery, 1 c
- Olive oil, 2 tbsp
- Nettle leaves, 4 to 6 c

You Will Do:

1. Bring a pot of water to a boil. As you wait, wash the nettles in a sink of cold water, using tongs. Place the nettles into the boiling water, making sure they stay submerged for two minutes. Pour them into a strainer and then run cold water over them to make them stop cooking.
2. Heat the oil in a pot and add in the onion and celery, cooking until the onion is translucent. Mix in the thyme.
3. Pour in the stock and salt. Let this come to a boil. Add in the potatoes.

4. As the potatoes cook, chop up the nettles and mix them into the soup. Simmer the soup for 30 minutes to an hour. Take the pot off of the heat and mix in the cream and buttermilk.

5. Use an immersion blender to make the soup smooth.

6. Add more pepper and salt if needed.

Creamy Nettle Soup

You Will Need:

- Pepper and salt
- Heavy cream, 2 c
- Juice of a lemon
- Zest of a lemon
- Nutmeg, .25 tsp
- Blanched nettle leaves, 4 c
- Chicken stock, 1 quart
- Butter, 2 tbsp
- Russet potatoes, 2
- Sliced and soaked leeks, 2
- Minced garlic, 2 to 3 cloves

You Will Do:

1. Start by mincing the garlic and set it to the side. Slice up the leeks and divide their rings out and soak them in some cold water to get rid of the dirt and said. Drain them. Peel and chop the potatoes.

2. Add some water to a large pot and let it come up to a boil. Add the nettles, working in batches, into the water for 30 seconds. Bring them out and put them straight into ice water. Drain them.

3. In a dry pot, add the garlic, leeks, and butter. Sauté them until fragrant and soft.

4. Add in the potatoes and stock. Simmer the mixture until the potatoes are fork-tender.

5. Add in the lemon zest and juice, nutmeg, and nettle. Bring everything back up to a simmer.

6. Set the pot off of the heat. Using an immersion blender, puree your soup until smooth. You can also use a blender and work in batches.

7. Mix in the pepper, salt, and heavy cream.

8. Serve with some sour cream or grated cheese.

Fiddlehead Soup

You Will Need:

- Pepper, .25 tsp
- Salt, .5 tsp
- Chopped chives, .25 c
- Heavy cream, 1 c
- Diced medium potato
- Bay leaf
- Chicken stock, 4 c
- Minced garlic, 1 clove
- Sliced spring onions, 3
- Butter, 2 tbsp
- Fresh ostrich fern fiddleheads, 1 pound

You Will Do:

1. Start by rinsing and cleaning the fiddleheads well with cold water to get rid of the brown skin. Drain them and set them to the side.
2. Add the butter to a pot and melt.
3. Add in the spring onions and cook them for about three minutes, or until they are tender and fragrant.

4. Add in the garlic and cook for another minute.

5. Pour the stock in and add in the bay leaf. Allow this to come up to a boil.

6. Add the potato and fiddleheads. Turn the heat down to a simmer and let it cook until all of the vegetables are tender. This will take about 15 minutes.

7. Add in the pepper, chives, salt, and heavy cream. Constantly stir the soup until it slightly thickens. Serve with a garnish of green onions.

Dandelion and Violet Lemonade

You Will Need:

- Water, 6 c
- Lemons, 10
- Cane sugar or honey, .5 c
- Fresh violet flowers, 2 c
- Fresh dandelion flowers, 2 c

You Will Do:

1. Begin by adding the water to a pot and bringing it to a boil.
2. Set the water off the heat and add in all of the whole flowers. Let them steep for 20 minutes.
3. Strain your flowers out of the water and add them to your compost, or dispose of them however you want.
4. Let the tea cool down to about lukewarm, and then add the sugar or honey.
5. Pour into a pitcher.
6. Juice the lemons and add this to the tea mixture. This will make the water change to a pretty pink color.
7. Stir together and adjust the sweetness to sit you. If needed, you can add more water to fill up the pitcher.
8. Refrigerate until it is cold before serving.

Ginger, Pineapple, and Purslane Smoothie

You Will Need:

- Diced ginger, .5 tsp
- Turmeric, .5 tsp
- Yogurt, .5 c
- Pineapple, 1 c
- Purslane, 1 c

You Will Do:

1. All you need to do is add everything to a blender and mix until combined. Enjoy.

Purple Dead Nettle Tea

You Will Need:

- Dried purple dead nettle, 1 tsp
- Boiling water, 1 c

You Will Do:

1. Steep the dried purple dead nettle in the water for about 10 minutes.
2. Serve with some honey.

Violet Syrup

You Will Need:

- Violet petals, 1 c
- Water, 1 c
- Sugar, 1 c

You Will Do:

1. Heat the water up to a boil and set it off of the heat. Add in the petals and cover. Let this sit for 24 hours.
2. Using a double boiler, heat up the water and petals and mix in the sugar.
3. Let this come to a boil, stirring often. Set it off of the heat and strain into a clean jar. This will keep in a fridge for six months.

Wild Violet Vinegar

You Will Need:

- White balsamic vinegar, 1 c
- Violet flowers, .5 c

You Will Do:

1. Add the flowers to a mason jar so that they fill the jar about halfway. Pour in the vinegar.
2. Cover with a lid. If the lid is metal, place a piece of parchment between the jar and lid so that the metal doesn't react with the vinegar.
3. Set the jar in a cool, dark place for a couple of weeks.
4. Use just like you would any other vinegar.

Dandelion Root Chai

You Will Need:

- Maple syrup or honey, to taste
- Milk
- Cold water, 3 c
- Whole allspice, 3
- Whole star anise
- Green cardamom pods, 4
- Whole cloves, 5
- Peppercorns, 1 tsp
- Cinnamon stick
- Ginger root, 1 inch
- Roasted dandelion root, 2 tbsp

You Will Do:

1. Start by mixing the dandelion root with the spices in a pot. Pour in the cold water.
2. Allow this to come to a boil and then reduce to a simmer. Let simmer for ten to 15 minutes. Set it off of the heat.
3. Strain the tea.

4. Add the tea to mugs, filling them about ¾ of the way full. Add in the milk to fill up the cup and sweeten to taste.

Fireweed Tea

You Will Need:

- Fresh fireweed leaves, 2 lbs

You Will Do:

1. Strip the good leaves off of the stalks and place them into a bowl.
2. Pick off some of the leaves and roll them between your palms. Set these rolls into a bowl.
3. Once all of the fireweed is rolled, place the lid on the bowl and place them out of sunlight. Let this sit for two to three days. Discard any leaves that mold.
4. Once the leaves are nearly black, place them in the sun to dry. You can also roast them for 20 minutes at 350.
5. Place this in a mason jar. You can now make some tea. Brew just like you would any tea.

Hibiscus Syrup

You Will Need:

- Sugar, .5 to 2 c
- Fresh hibiscus calyx, 20
- Water, 3.5 c

You Will Do:

1. Start by bringing the water to a boil.
2. Take the outer calyx off of the seed pod and place the rest into the pot.
3. After the water starts boiling, turn the heat down and simmer for 20 minutes.
4. Strain the hibiscus out of the water and place the water back in the pot. Add in the sugar and bring it back up to a boil. You can add as much sugar as you want. The more sugar you add, the sweeter it will be.
5. Boil the liquid until it reaches your desired consistency. The longer it boils, the thicker it gets. Five to ten minutes is typically enough.
6. Set the pot off the heat and let it cool. Pour into clean jars and keep it in the fridge for up to six months.

Wild Ramp Pesto

You Will Need:

- Lemon juice, 1 tbsp
- Olive oil, .5 to .75 c
- Sea salt, 2 tsp
- Wild ramp leaves, 6 handfuls

You Will Do:

1. Start by dividing your ramp leaves into the six different handfuls. Two handfuls will stay fresh while the other four will get blanched. This is to cut down on the intensity of the pesto because if you do all six handfuls fresh, you won't have to worry about vampires for the rest of your life. But, by all means, change up the ratio to your taste preference.
2. Add some water and salt to a pot and get it boiling. Working in small batches, blanch the ramp leaves for 20 seconds, or until they become bright green and wilt slightly. Then immediately place them in ice-cold water, and lay them out on a dishtowel to dry.
3. Once you have finished blanching the leaves that you want to blanch, place them along with the fresh ramp leaves into the food processor with the salt, lemon juice, and olive oil. Feel free to add anything else you would like in your pesto right now, such as pepper, walnuts, or parmesan.

4. Process the ingredients until smooth, adding extra oil as needed.

5. Taste and adjust the lemon juice and salt to taste.

Polish Fermented Mushrooms

You Will Need:

- Smashed garlic, 2 cloves
- Cracked pepper, 2 tsp
- Caraway seed, 1 tsp
- Dried dill, 1 tsp
- Crushed juniper berries, 6 to 10
- Pickling salt
- Cleaned mushrooms, 3 to 4 lbs

You Will Do:

1. Boil the mushrooms in some salted water for about five minutes. Drain and allow them to cool on some paper towels.
2. Combine the spices and herbs together in a bowl. Sprinkle a thin layer of pickling salt in the bottom of a non-reactive container. Place a layer of mushrooms on top. Sprinkle with some of the spice mixture. Add some more salt. Continue to layer like this until all of the mushrooms are in the container. Finish with a layer of salt.
3. Lay a plate on top of the mushrooms to weight them down. Place this is a dark, cool place for four days. Check on the mushrooms after the first day to make sure that the mushrooms are submerged in their brine. If they aren't, boil a pint of water with a couple of tablespoons of kosher salt. Once cooled, pour over the mushrooms.

4. After the four days have passed, place the mushrooms and their brine into a clean mason jar. Keep refrigerated. This will keep for a few months.

Garlic Mustard Pesto

You Will Need:

- Salt, .5 tsp
- Extra virgin olive oil, .3 to .5 c
- Grated parmesan, 1 c
- Garlic mustard leaves, 4 to 5 c
- Walnuts, almonds, or pine nuts .25 c

You Will Do:

1. Add the nuts to a food processor and pulse a few times so that they become large crumbs. Add in the parmesan and garlic mustard. Pulse until the leaves are minced, and everything is combined.
2. Continue to pulse the mixture as you slowly pour in the oil. You can eyeball how much you need. The mixture should become shiny and wet.
3. Add in the salt and pulse to mix it in.
4. Use however you would like.

Sorrel Sauce

You Will Need:

- Pepper and salt
- Vermouth or stock, 2 tbsp
- Sorrel leaves, .25 lb
- Unsalted butter, 3 tbsp
- Heavy cream, .66 c

You Will Do:

1. Slice the sorrel into very thin slices.
2. Pour the cream into a pot and let it come to a simmer. This keeps it from curdling once the acidic sorrel hits it.
3. In another pot, add the butter and sorrel. Cook, stirring often until it cooks down and the sorrel turns bright green.
4. Stir in the cream and bring the mixture to a light simmer. It is going to be thick, so you add in the stock or vermouth to help thin it out.
5. Add some pepper and salt, and enjoy.

Fennel Sauerkraut

You Will Need:

- Pickling spices, 3 tbsp
- Crushed juniper berries, 1 tbsp
- Pickling salt, 1.6 oz
- Shredded cabbage, 2.5 lbs
- Sliced fennel bulbs, 2.5 lbs

You Will Do:

1. Mix the cabbage and fennel together.
2. Place a layer of the veggies in the bottom of three-gallon crock about an inch thick. Sprinkle with salt and some pickling spices. Continue this until everything is in the crock. Place a weight on top of the kraut and place it in a dark, cool place.
3. Check on them the next day to make sure there is a brine that is covering everything. If not, boil some water and salt together and pour it into the crock.
4. Let this stay in the cool, dark place for at least a week, but up to a month.
5. Take the weight off and place the kraut in a quart-sized jar. Keep in the fridge.

Country Mustard

You Will Need:

- Salt, 2 tsp
- Water or white wine, .5 c
- Vinegar, 3 tbsp
- Mustard powder, .5 c
- Mustard seeds, 6 tbsp

You Will Do:

1. Grind up the mustard seeds in a coffee grinder or by hand. They don't have to be perfectly ground.
2. Pour the seeds into a bowl with the salt and mustard powder.
3. Pour in the vinegar and water or wine. Stir everything together. Pour into a jar and store in the fridge.

Ancient Roman Mustard

You Will Need:

- Salt, 2 tsp
- Red wine vinegar, .5 c
- Cold water, 1 c
- Chopped pine nuts, .5 c
- Chopped almonds, .5 c
- Mustard seeds, 1 c

You Will Do:

1. Grind up the mustard seeds in a coffee grinder or by hand. They don't have to be perfectly ground. Add the nuts and grind into a paste.
2. Pour the seeds and nuts into a bowl with the salt and water Mix together and let sit for ten minutes.
3. Pour in the vinegar. Stir everything together. Pour into a jar and store in the fridge.

Pickled Blueberries

You Will Need:

- Champagne or white vinegar, 1 c
- Sugar, 3 tbsp
- Salt, 1 tsp
- Blueberries, 1 pint

You Will Do:

1. Place the berries into a pint jar.
2. Boil the vinegar, sugar, and salt together. Pour this over the blueberries, leaving a bit of headspace.
3. Cover and refrigerate. These blueberries will keep for a year like this.

Pickled Fiddleheads

You Will Need:

- Dried thyme, 1 tsp
- Mustard seeds, 2 tsp
- Peppercorns, 2 tsp
- Bay leaves, 2
- Salt, .25 c
- Water, 1 qt
- Fiddleheads, 1 lb

You Will Do:

1. Trim the ends off the fiddlehead.
2. Bring a pot of water to a boil and add plenty of salt. Boil the fiddleheads for two minutes and then place them in ice water.
3. Dissolve a quarter cup of salt into a quart of water. Fill a jar ¾ of the way with some fiddleheads. Cover with the brine. Weigh them down so that the fiddleheads stay submerged.
4. Place in a cool, dark place for two weeks. If you get mold on the top of the brine, that is fine, just skim it off.
5. Divide the spices between the jars of fiddleheads and screw on the lid. Keep in the fridge.

Candied Angelica

You Will Need:

- Sugar, 1 c
- Water, 1 c
- Baking soda, .5 tsp
- Angelica, 1 lb

You Will Do:

1. Cut the angelica stems so that they fit into a jar. Boil a pot of water and add in the baking soda. Prepare a bowl of ice water. Boil the angelica for five minutes and then place it into the ice water.
2. Bring the water and sugar to a boil. Place the stems into jars and pour the hot syrup over them. Let them cool and then screw on the lid. Leave them at room temperature overnight.
3. The following day, pour the syrup into a pot. Let it boil and add in the stems. Boil for a couple of minutes and then everything back into the jar. Cool overnight again. Do this two more times.
4. After the last boil, place the stalks on a rack to cool and dry. Once room temperature, roll them in sugar and keep them in a jar.

Strawberry Dandelion Cake

You Will Need:

- Room temperature egg whites, 6
- Sugar, 1.5 c
- Softened butter, .75 c

Cake:

- Diced strawberries, 2 c
- Milk, .75 c
- Salt, .75 tsp
- Baking powder, 2 tsp
- All-purpose flour, 3 c
- Vanilla, 1 tbsp
- Room temperature egg

Dandelion Syrup:

- Vanilla, 1 tsp
- Honey, .25 c
- Dandelion tea, .5 c

Frosting:

- Milk, 2 tbsp

- Vanilla, 2 tsp
- Berry puree, .5 c
- Sifted powdered sugar, 5 c
- Softened butter, 10 tbsp

You Will Do:

1. Start by getting your oven to 350.
2. Get two 9" cake pans ready by greasing them with some butter and dusting them with some flour.
3. Start by taking the ¾ cup of butter, add the sugar and place them in the bowl of a stand mixer and cream them together. You can also use a hand mixer if you want.
4. Whisk the salt, baking powder, and flour together in a separate bowl.
5. Add the egg whites to a bowl and add in one whole egg. Whisk everything together.
6. Once the sugar and butter have been creamed, carefully add in the egg mixture a bit at a time. Mix thoroughly after each addition.
7. Mix in the vanilla.
8. Mix in a third of the flour mixture and then half of the milk. Add in another third of the flour and then the rest of the milk. Finally, mix in the last of the flour. Make sure everything is fully combined together, scraping down the sides of the bowl if you need to.
9. Stir in the strawberries.
10. Pour the batter into the prepared cake pans, dividing it as evenly as possible. Bake them for 30 to 35 minutes.

11. Once cooked through, cool them on wire racks as you finish making the topping.

12. For the dandelion syrup: Make a cup of dandelion tea by adding dried dandelion flowers into a tea ball and steep it in a half cup of hot water for ten minutes. Take the flowers out and stir in the honey.

13. For the frosting: Start by pureeing your berries in the blender.

14. Next, whip the butter until it is light and creamy.

15. Slowly sift in the powdered sugar until it is all mixed in.

16. Add in the berry puree and the vanilla.

17. You can add just enough milk to get it to the consistency that you like.

18. To put the cake together, place one of the cakes upside down on a plate and brush the top with half of the dandelion syrup. Top the cake with the frosting.

19. Place the second cake right side up onto the first cake and brush it with the rest of the dandelion syrup. Frost and decorate the cake the way you would like. You can even add some edible flowers or sliced strawberries as decorations.

Douglas Fir Poached Pear and Frangipane Tart

You Will Need:

Crust:

- Almond extract, .25 tsp
- Cold water, 3 tbsp
- Cold cubed butter, 10 tbsp
- Salt, .25 tsp
- Sugar, 2 tbsp
- All-purpose flour, 1.25 c

Frangipane Filling:

- Browned butter, 2 tbsp
- Almond extract, .25 tsp
- Slightly beaten eggs, 2
- Sugar, .25 c
- Almond meal, 1.25 c

Douglas Fir Poached Pears

- Cinnamon stick
- Douglas fir needles, .5 c
- Water, 4 c

- Sugar, 2 c
- Peeled whole pears, 2

You Will Do:

1. Start by bringing your oven to 375.
2. Using a food processor, combine the salt, sugar, and flour for the curst. Add in the almond extract, water, and cubed butter. Pulse until the mixture starts to look like wet sand.
3. Place the crust into a tart pan and use the back of a spoon to press the dough firmly into the bottom and up the sides. Let this sit in the freezer for 30 minutes.
4. Place some parchment paper over the crust and add pie weights, rice, or beans onto the paper. This helps the crust keep its shape while baking. Place the pan onto a baking sheet and bake for 20 minutes.
5. Check the crust to see if it looks dry. If it does, then the crust is done. If not, let it cook for another five minutes. Take the crust out and let it rest for five minutes before taking out the pie weight. Allow the crust to finish cooling as you fix the filling for the pie.
6. For the frangipane: Add the butt to a pot and brown it until it just turns golden and is fragrant. Set to the side.
7. Using a food processor, add the almond extract, eggs, sugar, and almond meal. As the processor is running, slowly add in the browned butter until well combined. Set to the side.

8. For the Douglas fir poached pears: in a pot, add in the sugar, cinnamon stick, fir needles, water, and peeled, whole pears. Allow this to come up to a simmer and cook for 20 minutes. You should be able to easily pierce the pear with a fork. Set off of the heat.

9. Take the pears out and place it on a cutting board. Clean off any needles that may be stuck to them. Strain the syrup of the needles and place it back into a pot and let it simmer until the syrup has reduced to two cups. Set to the side.

10. Once you can handle the pears, halve them lengthwise and remove the core and seeds. Trim the stem and blossom end of the pear, and then slice it crosswise into quarter-inch thick slices. Do your best to keep the pear form intact.

11. To assemble the tart, pour the frangipane into the crust, smoothing it out. Carefully pick up the pears, still in the pear shape, and place them onto the filling so that the stem endpoints to the middle. Carefully fan out the slices.

12. Bake this for 40 to 45 minutes, rotating the pan halfway through so that it evenly browns. Once the frangipane is golden, take it out and brush the pears with the syrup you made. Allow the tart to cool completely.

13. Slice and serve with some warmed syrup if you would like.

Wintergreen Ice Cream

You Will Need:

- Chopped semi-sweet chocolate, 4 oz
- Cornstarch, 3 tbsp
- Maple syrup, 3 tbsp
- Wintergreen extract, .5 tsp
- Wintergreen berries, 2 oz
- Sugar, .75 c
- Whole milk, 2 c
- Heavy cream, 2 c

You Will Do:

1. Reserve a quarter cup of the milk for use later. Heat the rest of the milk, wintergreen, sugar, and cream to a steaming point. Set if off the heat, cover, and let it steep for two hours. Once cool, pour into a lidded container and place it in the fridge. You can leave it here up to overnight.
2. Pour back into the pot and slowly heat it back up. Whisk the cornstarch and reserved milk together. Mix into the ice cream base. Mix in the syrup. Stirring frequently, bring the base back to steaming. Then stir constantly for eight to ten minutes.

3. Switch off the heat and cool. Mix in the wintergreen extract. Pour into an ice cream maker and follow its directions for setting up the ice cream.
4. Fold in the chocolate chips. Keep in the freezer until ready to serve.

Mulberry Sorbet

You Will Need:

- Cassis, 2 tbsp
- Mulberries, 5 c
- Water, 1 c
- Sugar, 1 c

You Will Do:

1. Start by removing the green stems off of the berries.
2. Add the water and sugar to a pot and let it come to a boil. Let this simmer for three to four minutes. Set off the heat and let it cool.
3. Add the berries to a blender and pour in the syrup you just made. Blend until smooth.
4. Press this through a sieve to get rid of seeds and stems.
5. Chill in the fridge for a few hours and then pour into an ice cream maker and follow its directions.

Gooseberry Sorbet

You Will Need:

- Sugar, .75 c – you can add more if needed
- Vodka, 3 tbsp
- Gooseberries, 8 c
- Water

You Will Do:

1. Add the berries to a pot and cover with water. Let this come to a boil and cook for two to three minutes.
2. Set off the heat and crush the berries to a pulp with a masher. Don't use any type of blender. Let this sit until room temperature. Pour through a sieve and refrigerate overnight.
3. Strain again, but place a piece of paper towel inside the strainer. You should get clear juice.
4. Take 1 ½ to 2 pints of the juice and sweeten it to taste. Mix in the vodka.
5. Use an ice cream maker and follow its directions to make the sorbet.

Wild Cranberry Sauce

You Will Need:

- Apple pie spice, 1 tsp
- Water, .25 c
- Maple syrup, 1 c
- Cranberries, 4 c
- Zest of an orange

You Will Do:

1. Start by adding everything in a pot and mix it together. Let this come to a boil, and simmer until the cranberries burst and let the liquid reduce slightly. This will take about 20 minutes.
2. Chill to allow it to thicken, and enjoy.

Paw Paw Ice Cream

You Will Need:

- Egg yolks, 5
- Vanilla extract, 1 tsp
- Sugar, 1 c
- Milk, 2 c
- Cream, 2 c
- Mashed pawpaws, 1.5 c

You Will Do:

1. Start by heating the milk, cream, and sugar together until steaming. Mix in the extract.
2. Beat the yolks together. While stirring the eggs, add in a ladle of the cream mixture. Do this one more time, and then pour back into the pot.
3. Stir and heat back up to a steaming point. Once thickened, it should coat your spoon. Turn off the heat and pour it into a bowl.
4. After the custard has cooled off, whisk in the pawpaw until combined. Pour into an ice cream maker and follow its directions. Keep frozen until ready to eat.

Black Walnut Snowball Cookies

You Will Need:

- Powdered sugar
- Stick of butter, cut into cubes
- Pinch of salt
- Grand Marnier, 1 tsp
- Orange flower water, 2 tsp
- Sugar, 2 tbsp
- Chopped black walnuts, 1 c
- All-purpose flour, 1 c

You Will Do:

1. Start by setting your oven to 300. Combine everything together, except for the powdered sugar. Mash everything together with your hands so that it looks like a lumpy meal.
2. Form into balls and place on a baking sheet. Place in the oven for 35 minutes. Let them cool for five minutes. When you can handle them, roll them in the powdered sugar. Once completely cooled, roll in the sugar once more.

CONCLUSION

Thank you for making it through to the end of the book, let's hope it was informative and able to provide you with all of the tools you need to achieve your goals whatever they may be.

The next step is to start gathering the tools that you need to start foraging. Foraging shouldn't be seen as something odd or unnatural. For a long time, foraging was the only way people could find food other than by hunting. If you choose to forage, it will help you save money, and it will give you the ability to consume food that likely has never been touched by pesticides or other chemicals.

It is important that you remember everything we have talked about when it comes to distinguishing plants. Your overall goal should be to make sure that you don't end up consuming a poisonous plant. It wouldn't hurt to make sure you have your phone with you as well as picture references to help you spot plants, especially when you are first starting out. Once you have foraged for a while, you will find it a lot easier to distinguish the good from the bad. Also, make sure you never take more than you need and don't cause severe damage to the environment. You want to make sure that what you take is able to grow back so that you will be able to return in a year or so and take more. You never want to kill a plant.

There is an endless number of ways to use these foraged plants. They make delicious meals, and they can be used to treat various ailments. The possibilities are endless.

Finally, if you found this book useful in any way, a review on Amazon is always appreciated!

www.ingramcontent.com/pod-product-compliance
Lightning Source LLC
Chambersburg PA
CBHW081746100526
44592CB00015B/2318